Garden Farming Series

Keeping Goats

Elisabeth Downing

PELHAM BOOKS

First published in Great Britain by
PELHAM BOOKS LTD
52 Bedford Square
London WC1B 3EF
March 1976
Second impression October 1976
Third impression October 1977
Fourth impression July 1979

ISBN 0 7207 0883 4

Filmset and printed in Great Britain by
BAS Printers Limited, Over Wallop, Hampshire

Contents

Preface

It only takes one generation away from the land for the knowledge gathered over thousands of years to be lost. How many of us have said, 'My grandmother used to keep a pig and we always had home-cured bacon when we stayed with her.' But poor dear grandmamma is now dead, and Cousin Emily can't remember for the life of her what she rubbed into the bacon anyway. No one bothered to write it down, as recipes were handed on from generation to generation by example and word-of-mouth.

The tradition of self-sufficiency has almost died in England; perhaps the Second World War tolled the death knell of a way of life which was the norm in the countryside. Ideas which were second nature to our grandparents are almost completely alien to the present generation.

Shortages and high costs of food have now made many of us who, even ten years ago, would not have dreamt of producing our own milk or even meat, now consider the question seriously. Most of us have not enough room in the garden nor the inclination to keep a cow, but what used to be known as the poor man's cow—the goat—can quite happily live with the family (as she invariably will) even if her official residence is at the bottom of the garden.

This book may, I hope, help the family who has always had a secret yearning for a goat and wishes to justify her existence in the family by helping her to earn her keep by supplying a wide range of dairy products.

Elisabeth C. M. Downing

Acknowledgments

A number of people have been particularly helpful in the writing of this book.

I would like especially to thank J. Crispin Clark BVMS, MRCVS for his great help and invaluable advice from contemporary practice in the writing of Chapter 7. He made time available for discussion and correction in the midst of his very busy life.

Appreciation is also due to Dulcie Asker for typing the manuscript with such humour and forbearance.

I would also like to thank the following goat keepers for their unstinted help and advice: Mr and Mrs Beckley; Mrs Rosemary Crawley; and Mrs Darrah. Also Mrs Wendy Fricker who encouraged and stimulated at all stages.

Credit should also be given to my husband for his work in annotating the diagrams.

Lastly, I would like to thank my family who were always available for criticism, correction and help, and who showed extreme forbearance in living on a diet of goat yoghurt for three months.

Conversion Table

Metric and Imperial Equivalents

Imperial	Metric	Metric	Imperial
1 inch	2.54 cm	1 cm	0.39 in.
1 foot	30.48 cm	1 cm	0.033 ft
1 yard	0.91 m	1 m	1.094 yds
1 mile	1.61 km	1 km	0.62 miles
1 sq yd	0.84 sq m	1 sq m	1.196 sq yds
1 cu yd	0.76 cu m	1 cu m	1.31 cu yds
1 pint	0.57 litre	1 litre	1.76 pints
1 gal	0.0056 cu m	1 cu m	219.97 gals
1 gal	4.55 litre	1 litre	0.22 gals
1 fl oz	28.4 ml	1 ml	0.035 fl oz
1 oz	28.35 g	1 g	0.035 oz
1 lb	0.45 kg	1 kg	2.20 lb
1 acre	0.405 hectare	1 hectare	2.47 acres
x°F	$\frac{5}{9}(x - 32)$°C	y°C	$(\frac{9}{5}y + 32)$°F

Milk

1 lb $= \frac{3}{4}$ pint
1 pint $= 1\frac{1}{3}$ lb

Metric abbreviations

cm	centimetre
m	metre
km	kilometre
ml	millilitre
g	gram
kg	kilogram

1 Buying Your Goat

Though she needs little land to support her, the goat can supply us with milk, yoghurt, cheese, butter, cream and even meat and skins for rugs if we are sufficiently hard hearted.

The weight of an adult nanny goat is, on average, about 150 pounds, depending on the breed, compared with the approximate 1150 pound weight of the Friesian cow. Obviously the cow is going to need vast quantities of high quality foods which are expensive to buy and impractical to grow in the average garden. The humble goat however, being so much smaller, consumes the equivalent of 25 pounds daily of browsings, rough vegetation and possibly grass, compared with well over 100 pounds weight of good quality grass, besides the additional dairy cake, needed by the milking cow. From this we can see that with a little thought and effort, we can justify the purchase of a goat to supplement the family diet and she will also more than justify her existence with her charms and whims.

We must primarily like the goat—it is no good setting out with the idea of being self-sufficient in dairy products while not being able to stand the sight and sound of the creature. The addition of a goat to the family is going to need thought and organization.

We must consider—are we going to have the inclination to get up each morning to milk her, come rain or shine, and once again in the evening? She will need exercise if she doesn't take to the tether—the Anglo-Nubian breed for example is well known for disliking any personal restraint. She will need companionship, human or otherwise, if she is to flourish and give her best. She will need attention 365 days a year, year in, year out. Admittedly, we can take her on holiday with us in the back of the car, and she will provide fresh milk daily into the bargain, but she will also turn the back of the car into a wreck within hours, with her browsing on the upholstery and the inevitable deep litter underfoot. On the other hand, a kind and knowledgeable neighbour may take her in for a day or two, or even come to your house to look after her if you go away for a few days.

The goat must have the approval of the family from the most senior member down to the family cat or mouse. She will want to share the kitchen, and even the living-room with her human friends, however palatial her own dwelling may be, and her hay and straw will invariably blow up the garden path when visitors are about to arrive, ruining all your attempts at a tidy garden. The manure from her house will have to be sited at a tactful distance from the dining-room window if you are not to be overcome by flies in August.

Having decided that you really cannot exist without the extra member of the family, it must be decided how to acquire this paragon.

There are several ways to set about obtaining a goat. The most facile is to have a charming kid, but two weeks old, given to you with the promise that before long you will be living in a land flowing with milk and cream. This is all very well, but she may not be the breed you like, or the colour, you may not know how prolific her dam is, she may very well not be pedigreed and you may wish to own and breed pure-bred, pedigree goats. Above all, unless one is experienced, it is very difficult to judge the virtues of a goat when she is so young. The appealing aspect of the young kid may well overwhelm the obvious drawbacks—an ugly, poorly bred, potentially low yielding goat. From this it can be seen that no little thought must be put into her choice and acquisition.

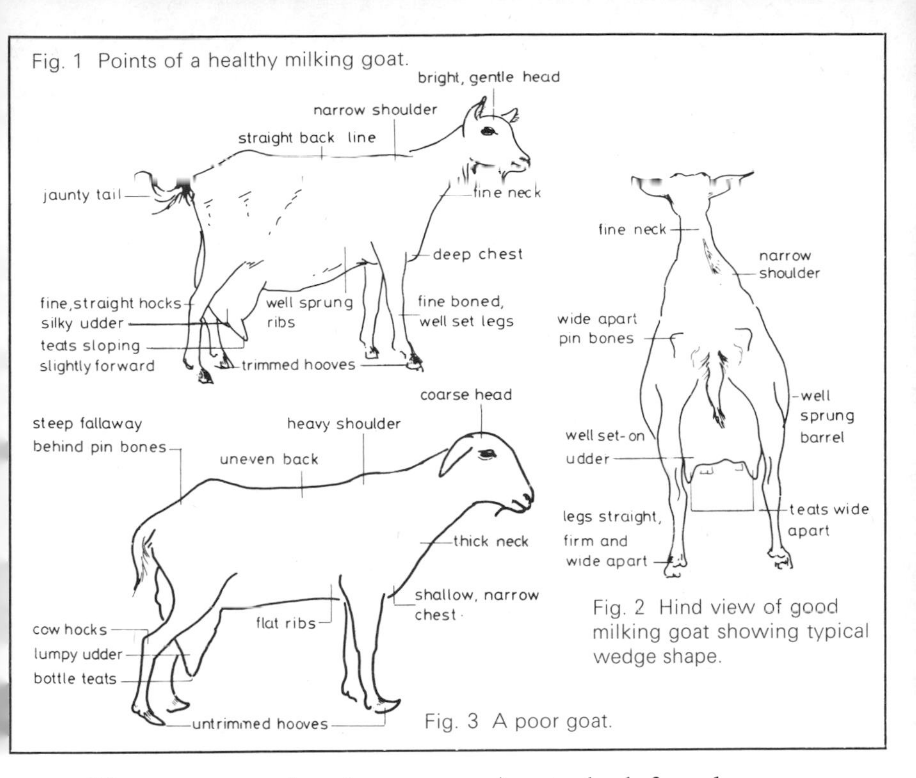

Fig. 1 Points of a healthy milking goat.

Fig. 2 Hind view of good milking goat showing typical wedge shape.

Fig. 3 A poor goat.

Figures 1, 2 and 3 give some points to look for when choosing your goat.

Initially, it is a good plan to visit some goat shows or the goat tent at the local annual county agricultural show. The goat fraternity is proverbially friendly and generous, and only too willing to extol the virtues of the various breeds and, possibly, even their respective drawbacks. Into the bargain, one's eye can be attuned to the shape and aspect of the well-nurtured, well-bred animal. There is a vast difference in breeding and looks between the show animal, which very often produces quantities of milk besides being pleasing to the eye, and the ill-bred, coarse-boned beast tethered (come rain or fine) down the road, which barely produces enough milk to justify her existence. This is not to say that all non-pedigree animals cannot produce well, but the animal of known parentage starts off at an advantage in that you know what she *should* produce given a reasonable chance, and what she may look like as an adult into the bargain. The well-bred goat will also never cease to be pleasing to the eye.

A visit to several breeders to discuss the virtues of the different breeds, types of housing, methods of feeding and general methods of management will help to give an idea of how to approach the subject with your own set-up in mind.

There follows a very brief outline of the main breeds found commonly in England. The word 'British' used in conjunction with the breed name is used to show that the animal is cross bred to a certain extent.

Saanan—A breed which originated in Switzerland. It is generally larger bodied than the other breeds with possibly a slightly concave or 'dished' profile and small prick ears. It is white in colour with a fine short coat, though black spots on the udder, ears, eyes and nose are sometimes found. One of the obvious advantages of the breed is the fact that heavy yields are common and lactations long. They also have a very placid temperament. See Fig. 4.

Toggenburg—Another breed which originated in Switzerland. It is generally smaller than the Saanan but still has the 'dished' face and prick ears. It often bears long hairs along the back and down the hind legs. Brown or fawn in colour, it has white markings. White stripes run from the muzzle to the eyes and round the edges of the ears. White is also found on the legs from the hocks and knees downwards and round the tail. This breed *may* not milk quite as well as the Saanan but it is very affectionate and, being smaller, may require less food than most.

British Alpine—This breed was developed in Britain and looks very like a black Toggenburg with the same

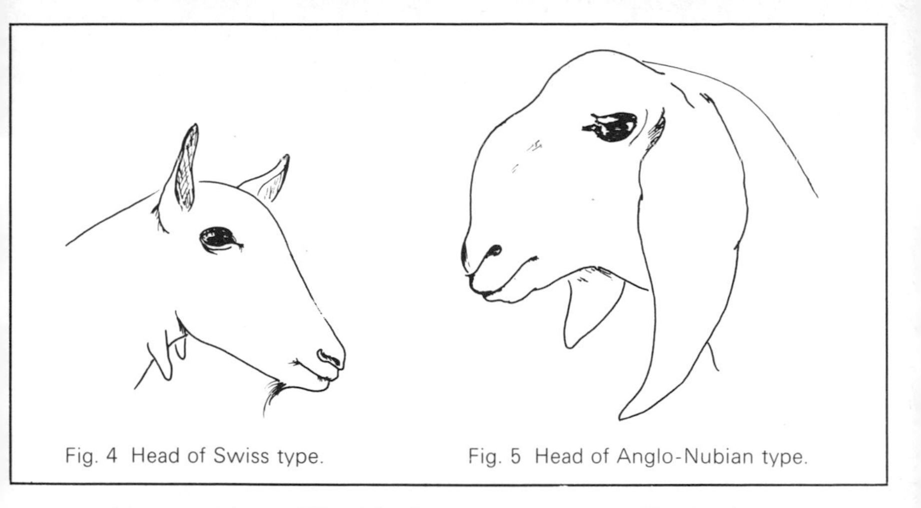

Fig. 4 Head of Swiss type.

Fig. 5 Head of Anglo-Nubian type.

white markings. The black coat may attract flies in the summer and the udders of this breed are sometimes rather unwieldy.

Anglo-Nubian—This breed has been developed from a combination of Indian, Nubian, and other eastern breeds with a fair sprinkling of native British blood. The distinctive features of a Roman nose and large lop ears (see Fig. 5) are highly prized by some. Colours range from black to white, while red, roan, sometimes in conjunction with black mottling, are commonly found. This breed is well able to stand extremes of temperature. It is extremely vociferous and objects strongly to being tethered. However, the butter-fat percentage of the milk is higher than that of other breeds as well as the solids-not-fat, but the milk production is not so heavy as that of the Saanan and the lactation possibly shorter. It is well suited to free-range management as the udder is generally better hung than other breeds.

Golden Guernsey—This breed is becoming more commonly known in Britain since the register for pure bred stock was recently opened. The breed originated in Guernsey and like the Guernsey cow is golden to a greater or lesser extent. It is a small, fine-boned animal generally.

Though the Anglo-Nubian does not carry tassels—the soft hair-covered appendages which hang under the neck—they may or may not be found on the other breeds. No one knows their use or origin. They are sometimes known as toggles or wattles.

Horned and naturally hornless animals occur in all breeds of goats. Disbudding naturally horned animals can be performed soon after birth (see page 75).

Having selected the likely breed, contact breeders of your chosen variety and find out if they have any stock for sale. Breeders of pedigree stock will be able to show you the parents of any animal they have for sale and also show you their milk records—either those kept officially and checked by the Milk Marketing Board, or simple ones kept for their own information and use. The exact age of the animal will be known from her registered papers. Breeders with pedigree animals for sale have a reputation to keep which they will not want to lose by selling poor stock.

Advertisements in the local paper under 'Animals For Sale' can be followed up, though from experience I know that many miles can be covered finding the prospective sellers, only to discover indifferent stock which sounded so convincingly suitable in the advertisement.

The age of the animal is another consideration. Buying a young kid will involve the acquisition of milk or milk substitute to feed the animal till she is old enough to subsist on adult rations. But rearing a young kid is not a task to be undertaken lightly, especially if you have no previous experience of doing so. Until some experience has been gained in the goat world in general, this may prove one of the most expensive methods of acquiring a goat both in feed costs and labour. Faulty feeding while the kid is still on the bottle, can even prove fatal in some instances; far better to leave it to the breeder. Of course, once it is weaned you will have an animal which is well accustomed to your methods of management and feeding. Some goats are notoriously conservative in their feeding habits, this will give you plenty of time to give her the widest possible range of foods both succulents and concentrated grain foods. This animal will also have to be mated at about eighteen months of age and still no milk will be flowing till she possibly kids five months later. A long time to wait!

On the other hand buying a ready-weaned, strong, young goatling of a year old or more will give you only a year to wait for her to start producing. A simpler method than the above is to buy the ready-mated goatling. Although she will only have a relatively short time to get used to her new home before she kids, it won't be long before she may be making you a handsome return.

The ready-kidded goat is considered by some to be the simplest way. This, I feel, is misguided. You will be milking straight away but she may well suffer a sharp drop in milk production owing to the change in management and diet, however small. Milking animals, be they goats or cows, are difficult to get back to previous production levels if they suffer any kind of change in management, regardless of quality.

Buying an older goat may appeal to some but she does not generally take kindly to changes in management and you may well have difficulties in getting to know her ways and food fads, compared with the younger goat who is keen to learn and less conservative in her ways. Another consideration is the fact that she may well have had disease in her udder at some time which can adversely affect the flavour of her milk at best, and at worst can affect milk production to a greater or lesser degree depending on the cause of the mastitis.

For thousands of years goats have been bred

specifically for milk production resulting in an animal which may produce far more milk than is needed for its young. This means that very often a female goat which has not even seen a male goat may start to lactate any time after maturity (not to be confused with 'cloudburst' (see page 103)). This is most likely to happen in spring time as the days lengthen and there is an abundance of succulent feed available. Very often the maiden milker does not have as big an udder as the goat which has been mated, followed by pregnancy, kidding and normal lactation. Her level of milk production will usually be lower.

When buying an in-milk goat, some proof of kidding would avoid the pitfall of buying a possible maiden milker. The owner of pedigree stock would have a certificate of service available and also the registration papers of any resulting kids. There is no reason, however, why a lactating maiden milker should not be mated normally if she comes into season.

Generally, buying an in-kid goatling (providing she is not less than fifteen months old at mating) would prove the most satisfactory method of acquiring your first goat.

Having decided on the breed, the breeder, and the approximate age of goat to buy, visit the prospective seller and be ready to take note of several points. Watch the way the owner treats the animals. Well-handled goats are interested, confident and amenable to the advances of even a stranger. Ask to see possibly the dam, and even grand-dam if available, and note how they have stood up to several kiddings. It is no good buying an animal whose udder will be in need of support and trails on the ground after the strain of a couple of lactations. Discuss the temperament of close relatives of the goat you may buy—do they stand quietly when being milked and do they mind being tethered? It is always useful to have an animal which will accept the tether even if you intend to graze her in a fenced paddock most of the time. Find out if she is an inveterate fence jumper or a positive 'Houdini' at getting out of her collar! Discuss the methods of feeding the animal has been used to, the housing and the general day-to-day routine. Ask to see the pedigree if the animal is registered with the British Goat Society. If she has been covered (i.e. mated) ask for the certificate of service so that you will be able to register the progeny if you wish.

It is always advisable to see the stock before you agree to purchase. An animal may sound superb on paper or even over the telephone, but when you see her you may just not 'take' to her. Having an animal which just does not appeal to you is getting off to a bad start! Buy the best stock you can afford, though very often pedigree stock even from officially milk recorded animals may cost very little more than one of poor quality with unknown parentage and no records.

The transportation of goats is relatively simple; they travel happily in the back of a car (*not* in the boot). An estate car is ideal but not a necessity. The floor can be lined with a split plastic fertilizer bag, opened out to cover the floor, with a comfortable layer of straw on top to mop up the inevitable urine; a nervous animal urinates more frequently than normal. The goat will very soon lie down and if there is some really palatable hay for her to pick at, she may very well be eating within minutes of getting into the car. If an animal is accustomed to car riding it will jump in on its own, but you may need to lift it by putting your arms firmly around the body (see Fig. 6), reassuring it with your voice all the time. A metal grid, such as dog owners employ, will keep the goat within bounds and prevent the upholstery from damage.

Fig. 6 Lifting the goat. Fig. 7 Introducing the newcomer.

The sound of your voice during the journey, besides reassuring the animal, will accustom it to the new surroundings. Goats, being herd animals, like to have a leader. The sooner it recognizes its new owner as herd leader the quicker it will settle down.

On getting home gently lift the animal out of the car; it may be lying down and after a journey animals are very often loth to get to their feet. Lead her to her house, still talking to her continuously and encouragingly. Don't pull. No animal appreciates this type of handling and most will resist strongly by bracing all four legs against the puller! A wide collar around the neck held up while you walk at the animal's shoulder saying, 'Walk on' firmly, and praising any forward movement at all will soon get a response (see page 114). Put the animal in her prepared house; a warm well-littered bed, a hay rack filled with good hay and fresh, preferably warmed, water. Do not underestimate the power of a firm reassuring voice when handling goats, or any other animal, for that matter. Confidence will rapidly build up between you and even if you feel a little shy at first to be heard talking to your goat, it may very well stand you in good stead in the case of an emergency when dealing with an extremely frightened animal. Hearing your calm voice she will associate this with pleasure and relax.

Leave the animal for an hour or so to sniff round her new house. Resist the temptation to peep in at her every few minutes and don't invite all the neighbours to inspect the new animal within hours. She needs peace and quiet.

A goat in strange surroundings may well look very different. Her coat may lose that soft smooth look and develop a rough feel. She may lift the hair along her back in her anxiety, urinate frequently and stand with her head slightly lowered and her ears back. You will have already found out from the previous owner how she has been fed, but when preparing her first evening feed of what are known as concentrates (grain feed) cut down a little on the quantity. She will be feeling strange and may not want to finish a full feed. A smaller feed than normal will sharpen her appetite for the next day. If she seems lonely don't be frightened to remain with her, talking all the while. Run your hand through the feed. Goats have inquisitive natures and her response will be to look in the strange bucket and pick at the food. If she shows no interest, take the bucket away and dispose of the food elsewhere. Don't offer her the same food next day, but offer some more fresh from the sack. Tempt her with a few succulent green leaves and fresh hay. It's a good plan to give her warmed water especially if the weather is cool and this will encourage her to drink. She may not lie down for some time but when she does, and you see her chewing the cud, you may be sure that she has settled in well.

This is the time to start introducing her to the rest of the family—not forgetting any other animals you may have. This needs to be done gradually over the next few days. Dogs, especially, often feel very jealous of the new pet and show their feelings by being over-bumptious or even trying to chivy her. If the introduction is done tactfully, talking to the dog rather than the goat, the dog will quickly accept the new member of the family and before long they will be firm friends.

Movement of animals from one home to another can cause stress and it may not be evident for three days or so. Of course the goat may settle in perfectly happily with no ill effects at all, on the other hand changes in diet may well cause her faeces to become a little loose. Her coat may be rougher and she may lose her appetite. A light diet, cutting down on the grain ration, but plenty of sweet hay and browsings (see page 42) will soon settle this. After the third day has passed with little or no adverse physical upset, you may be sure that your goat has settled in well. She will soon bleat on hearing your footsteps or voice, and will be looking bright-eyed and spending much of the time chewing the cud.

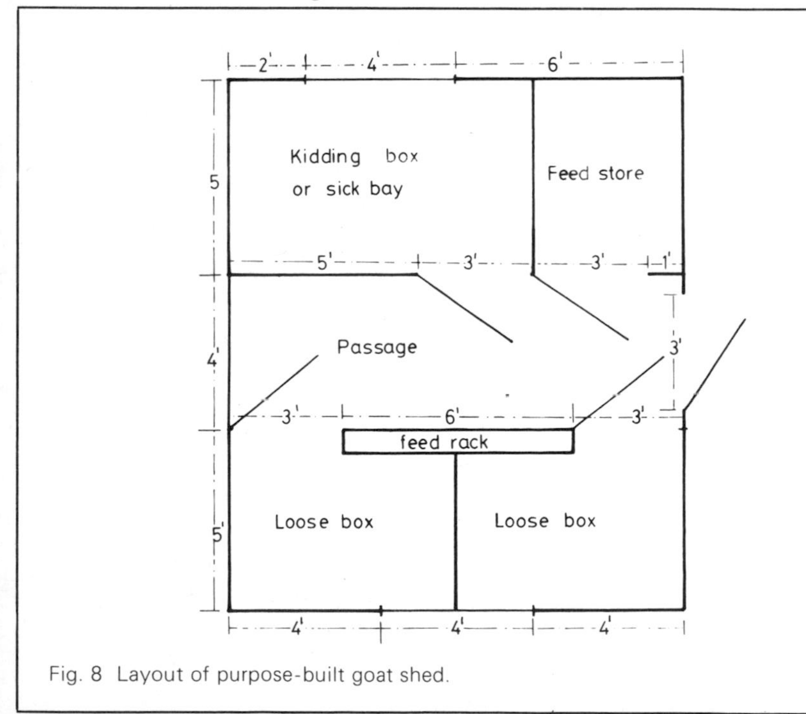

Fig. 8 Layout of purpose-built goat shed.

2 Housing

Goats much appreciate shelter. In fact it is essential to their health and happiness. How often have we seen the dejected, bedraggled animal, tethered on a windswept grass verge, with her tail into the wind and rain, hanging her head in misery? This is the quickest way to bring down the wrath of every goat keeper upon your head, and rightly so. A goat kept this way will likely as not become unprofitable and sick very quickly.

This does not mean to say that she needs a palace to live in. Far from it. Ideally she needs a well-ventilated, airy shelter, which is warm and dry underfoot. The nearer it is to human habitation the happier she will be. Goats are herd animals and like company, human and otherwise. Outside the back door is not necessarily the most convenient siting for us or the goat. Somewhere reasonably well drained, sheltered from the north and the prevailing wind, and into which the sun can reach for as much of the day as possible is ideal.

You may be lucky in having accommodation that can be easily adapted to the goat's needs. Basically she will need about 24 square feet of floor space. An old stable, outhouse, or even a chicken house (the type that is wheeled and meant to accommodate about twenty-five birds) will do. A garden shed may also be quite suitable. Figure 8 shows the layout of a purpose-built goat shed.

Existing buildings not being available, a suitable shed can be erected quite cheaply using old packing cases or even secondhand timber. Be guided, within reason, by the dimensions of the material available. For example, an old window frame 3 feet by 2 feet can readily be adapted to make a door. Old prefabricated house floors come in suitable sizes for roofs and are well insulated into the bargain. They will only need a skin of mineralized roofing felt to make them waterproof. Corrugated iron can be used for roofing, but with reservations. It is noisy when it rains, and also requires careful insulation to prevent condensation. It is also liable to rust away at the most inopportune times unless regularly tarred each year or so.

I have seen a goat shed, which accommodated several goatlings, made from a packing case used initially for transporting valuable machinery. It was already lined and measured 7 feet wide by 9 feet long, being 5 feet high. It cost £2. A doorway was cut in one end, the roof clad in mineralized felt, and at the other end a piece of corrugated asbestos leant up to the roof to make a lean-to and additional airy shelter.

For anyone lucky enough to live near forestry plantations, thinnings can be utilised for the internal divisions and fittings.

For those who are more ambitious, a reasonably cheap building may be erected which can consist of two separate living compartments with a further one for milking. The internal divisions can be made of weather-boarding or any other available timber. These are put up 3 feet high, with a further 2 feet on top consisting of bars (forestry thinnings will do here) 2 inches apart, so that the animals may see each other. The whole of the stall divisions can be made of spaced slats if necessary but this makes the stalls less snug.

All the timbers used need to be well planed and creosoted before use to withstand the ravages of hoof, mouth, insects and the elements.

Care must be taken over the ventilation. The object is to change the air frequently enough to keep the atmosphere fresh, without allowing draughts to develop. Both draughts and a stuffy atmosphere can result in the respiration of the animals being affected, which may lead, at worst, to pneumonia and can certainly contribute to a reduction in general health. A damp, muggy atmosphere providing a breeding ground for harmful bacteria and moulds is most unhealthy.

When considering ventilation, one must bear in mind the type of goat one is hoping to keep. If the animal is to be stall-fed for much of the time, she will need a well-insulated house which will keep her draught-free and warm. This obviously means that if she is to be healthy the air must be fresh. This will entail designing a ventilation system which is more complicated than that required for the outdoor foraging goat whose needs for shelter are intermittent. Windows of the hopper type, situated under the eaves, can be opened and closed as necessary, and will be least likely to cause draughts (see Fig. 9). Ventilation in the roof can be adjustable and will prove adequate for most goat sheds. A ventilation outlet, 18 inches by 1 foot, controlled by a sliding shutter at one's height on the wall opposite the door can provide extra ventilation in really hot weather.

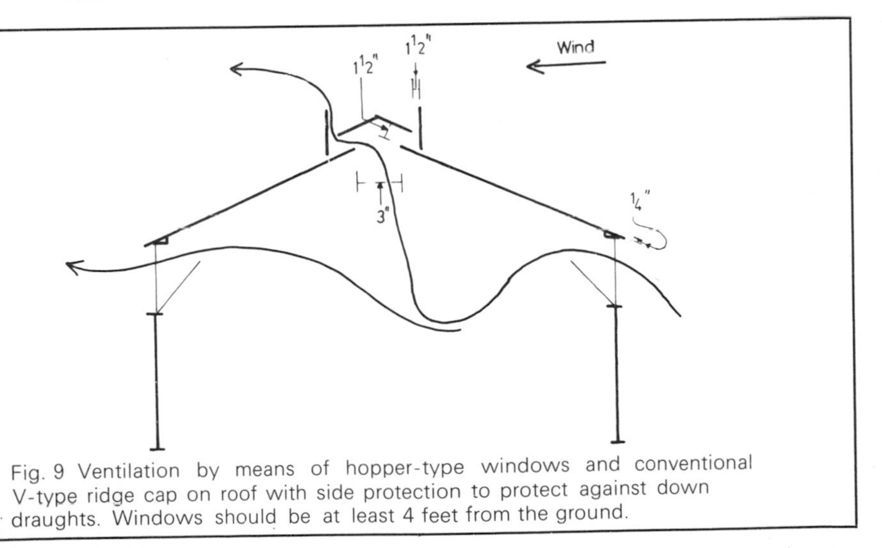

Fig. 9 Ventilation by means of hopper-type windows and conventional V-type ridge cap on roof with side protection to protect against down draughts. Windows should be at least 4 feet from the ground.

A simple form of shelter for the basically outdoor goat can be made from straw bales. She will spend much of her time browsing on roughage and will generate her own warmth (see page 45). The straw bales can be built up to form an open-sided shed. The single span roof should have an adequate overhang to keep rain from rotting the bales. Wheat bales are more suitable than barley or oat and should be baled with plastic bale-twine which will not rot easily in damp conditions. Chain link or heavy-gauge wire-netting erected 6–9 inches away from the bales will protect them from the ravages of the goat's mouth and hooves. This type of shelter will, with luck, last up to a dozen years. Take care to select a dry, high spot, an earth floor will then be adequate.

Flooring in the more elaborate house needs some thought. Concrete is reasonably cheap but very cold unless insulated. A concrete sub-floor can be laid 3 inches thick, covered by a bituminous layer, with $1\frac{1}{2}$ inches of concrete on top. This will provide some insulation against cold and damp, it will also be reasonably easy to keep clean. Breeze blocks with a layer of concrete on top are also warmer than plain concrete, but however concrete is employed it is cold for the goat. On the other hand, a small concrete area is ideal for helping to keep the goat's hooves in good order. Failing this, a stone or rock about 2 feet square and up to 18 inches high for the goat to jump up on may, besides being a useful asset for the alleviation of boredom, keep the hooves in hard, healthy condition.

A floor of rammed chalk 6 inches thick will provide a far warmer floor than concrete but it is not so easy to keep clean. A compromise could be a concrete-floored milking area (which can be easily swilled down after milking) and a rammed-chalk or earth floor in the living or sleeping area, provided the soil is light enough to allow this.

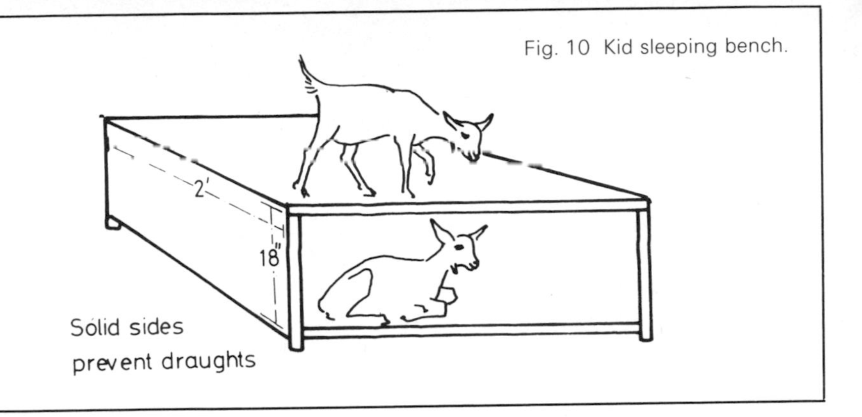

Fig. 10 Kid sleeping bench.

Goats enjoy their creature comforts, and, being natural climbers, will enjoy jumping onto a raised wooden area to sleep. An old door raised 8–12 inches off the floor on some bricks is adequate for one or two goats. Unless the weather is cold, they will need no litter on this area. The board need not be secured so that when cleaning the droppings off each morning, it can be lifted up and the manure scraped into a bucket or a plastic bag and then tipped into a barrow outside. Kids will very often like to sleep under a board, and in this case it needs to be about 18 inches from the ground (see Fig. 10). A more substantial bench should be erected which will stand up to being a play area as well.

Drainage can raise problems if a concrete floor is used. The obvious solution would be to slope the floor towards the doorway. However, this can result in an accumulation of liquid where the traffic is most dense. In practice, the drainage should be towards the back of the house. A floor with a front to back fall of 1:60 (in a box 5 feet wide this will mean a 1 inch fall) will suffice (see Fig. 11). At the back wall a narrow channel, wide enough to accommodate a yard broom sideways (and not deeper than 3 inches) can lead to a trap incorporating a sediment chamber with a grating to collect surface solids. A simple drainage hole will create draughts and allow vermin to come in. Outside, a small soakaway can be dug to accommodate the liquid.

Goats' solid excreta are normally passed in a dry pelleted form. This means there is far less mess than with other larger farm animals. The drainage in the sleeping quarters, especially if a deep litter system is employed on earth or chalk, can be minimal if the soil under the building is basically sandy and well drained. Problems do arise, however, with heavier soils under wet conditions, and in this situation rammed chalk will be more suitable than earth.

Goats enjoy being able to pick over food with interest and care. You only have to watch how she walks along taking a leaf here, a seed head there, or a choice morsel way up above her head. She does not, however hungry she is, stolidly munch through a yard-wide swathe of grass leaving a bare track behind her as a hungry bullock might do. When considering making the hay and fodder rack, this must be borne in mind. A hay net, then, may be filled with the choicest hay incorporating all her favourite plants. She will poke her nose in, sniff, possibly sneeze and retire with one blade of hay not even persevering to get at the dried nettles you placed in the net for her. You might get the impression on returning an hour later, that the hay was either not up to standard or that she'd had an ample sufficiency. In reality, however, it was not placed as she would have liked it. On the other hand, hay thrown even on the cleanest floor, she will merely pick over and perhaps even munch greedily for a few minutes, but her questing nose will then spread the valuable food all over the floor and as she reaches up for that extra succulent morsel, she will tread on the remaining hay. This will then have no more use as food as far as the goat is concerned. From this it can be seen that some device which allows as much hay and fodder to be exposed to her selective nose as possible

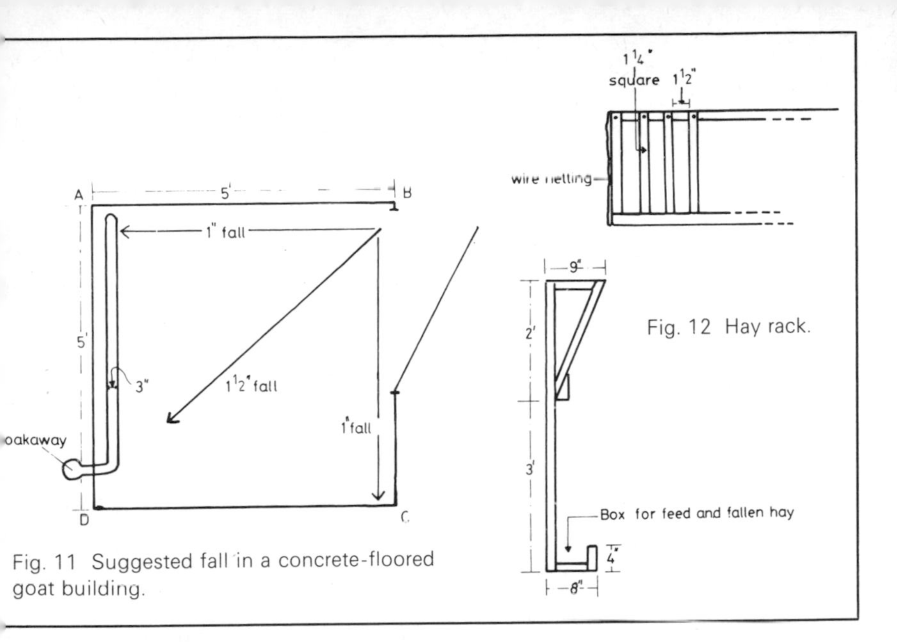

Fig. 11 Suggested fall in a concrete-floored goat building.

Fig. 12 Hay rack.

(without risk of wastage) is badly needed.

Over the years goat keepers have found that the hay rack has proved ideal (see Fig. 12). Wooden slats about 1¼ inches square placed 1½ inches apart would be ideal for this. Old chicken perching with bevelled edges can be used. Forestry thinnings 1½ inches in diameter are also suitable. A more sophisticated type of rack may be used especially where more than one goat is housed in the stall.

When feeding concentrates, animals must be individually rationed to avoid over-eating by the greedy goat. A rack which combines the ability to ration if necessary and also allows full access most of the time is ideal. This consists of a yolk through which the goat's head can pass and into which it can be secured by means of a hinged flap when necessary. This rack can also accommodate kale, cabbage, leaves, hogweed and other delicacies your goat appreciates. Branches of her favourite trees are best tied firmly in a bunch and suspended from a nail high enough to keep all the leaves off the ground or they may be trodden on

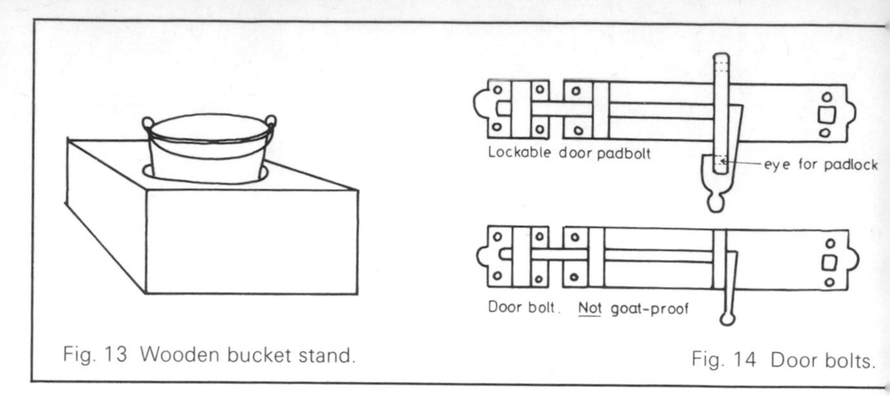

Fig. 13 Wooden bucket stand.

Fig. 14 Door bolts.

and soiled. A foot rest tacked 4 inches from the ground, below the nail, will help kids to reach the leaves. Small iron racks on the lines of those found in stables can be obtained from goat equipment specialists, but these are expensive.

Goats can be trained to drink from automatic drinkers. These are not easy to keep scrupulously clean. A far cheaper and foolproof method of providing water is in a plastic bucket. This can either be suspended from the wall within a substantial wire loop (this can be time-consuming when taking it down and then replacing it), or, ideally, the bucket can rest in a wooden bucket stand as shown in Fig. 13. The bucket can easily be removed for frequent cleaning and refilling. I like to scrub the buckets out thoroughly, daily; I empty out and refill the buckets each time the goats are visited using warm water in the cooler weather. White plastic will show up dirt and algae growth and encourage thorough scrubbing. A plastic scrubbing brush hung near the tap will give no excuse to shirk this responsibility. A water tap in the goat shed is unwise. Left on inadvertently, or even dripping, it will flood the goats out over-night. An inquisitive animal may even turn a tap on unwittingly.

Doors on the goat shed must be goat-proof. A bolt on the outside is suitable though inconvenient and time consuming to use. We have all heard of the goat who can untie knots and unfasten simple latches.

Start off with a really substantial one which, though it takes time to unfasten, will save the heartache of tending the sick goat who has let herself out of her shed and either overfilled herself with bloat-making brassicas from the garden, or helped herself to a week's supply of concentrates from the meal house. (See Fig. 14.)

The goat-house doorway needs to be wide enough to accommodate a wheelbarrow which will facilitate cleaning out and thus save time. The hinges on the door need to be substantially made in order to withstand the assaults of goat and wheelbarrow.

A separate larger box for kidding is ideal but not essential. It should have no interior fittings and be constructed to enable thorough cleansing between kiddings. Here, a concrete floor would facilitate cleaning but ensure that enough litter is available for the goat to be able to lie on a really deep bed for warmth and comfort during kidding.

A shed that is thoroughly goat-proof is essential for the storage of foodstuffs. Plastic dustbins will take, roughly, ½ cwt of most foods. These are dry and vermin proof but goats soon learn to lift off the lids or even turn them over to get to the contents. Dustbins with screw-on lids are obtainable from some iron-mongers and would suit. The door to the store must be securely fastened *at all times*. A bolt on both sides of the door will be safest. When preparing food it is easy for a goat to follow and steal a ready-prepared feed if the door is not shut behind the feeder.

Hay and straw do not appreciate being stored in close proximity to concrete floors and walls. Blocks of wood or boards raised on bricks will allow plenty of air to pass under the bales. Don't store the bales close to the walls but leave a foot between the straw and the concrete wall. Hay and straw can be stored outdoors but far less waste will result if it can be kept under cover. A simple shelter can be built from four corner posts with rafters supporting sloping corrugated iron sheets.

The goat paddock needs adequate fencing if the animal is to be confined efficiently. Four foot chain-link fencing of 2 inch 12½ gauge or 3 inch 10½ gauge, with posts at 6 foot centres is reasonable, provided that two or three strained support wires are adequately attached (see Fig. 15). An extra tight wire can be erected 9 inches above the chain link to deter the show-jumping goatling (the milker is not so likely to jump very high). The posts must be well dug-in as they ask to be used as rubbing posts. An advantage of chain-link fencing is that a tuck can be taken in or let out, and then relocked on uneven ground. Wire netting, however thick the gauge, does not stand up to the ravages of the goat's hooves, especially if there is a succulent hedge behind it.

Electric fencing is suitable provided that the animals are taught that they will experience a mild shock when it is touched. Wet the face, chest and neck of the goat to increase the shock and her conductivity, offer her the most succulent titbit you can find. Lead her to the fence holding the food on the opposite side so that she will have to touch the fence in order to get at the food. The resulting shock will, we hope, deter but not damage her. The electrified wire must be tight to be efficient and three wires will be needed. The fencer unit is expensive new, but secondhand ones can often be found at farm sales quite cheaply. The makers of the unit supply adequate instructions for erection and use. Remember that twigs or grass, particularly if wet from dew or rain, touching the wires will effectively reduce the shock by causing the current to earth. This needs to be checked several times a day.

Paled fencing should have a 3 inch gap between the

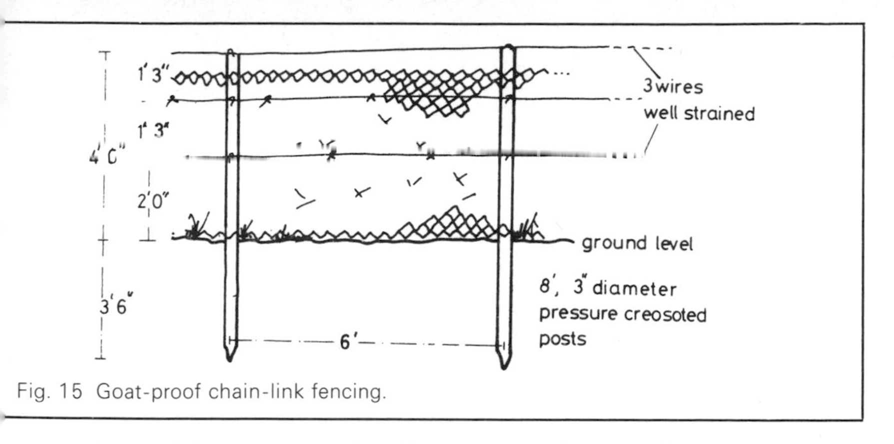

Fig. 15 Goat-proof chain-link fencing.

pales, while a post and rail fence can have rails 7–8 inches apart to avoid kids hanging themselves.

Barbed wire *must not* be used under any circumstances.

The only remotely efficient goat hedge is blackthorn and this needs to be well established and very thick. Remember that the thorns on this bush often cause an infection if allowed to enter the flesh of a human. Hawthorn may appear adequate, but it is unable to withstand the continual browsing of the goat, as it is extremely palatable.

The alternative to fencing the goat is some sort of physical restraint. As mentioned earlier, the Anglo-Nubian breed generally do not take kindly to tethering. However, most other adult goats will acquiesce with fair grace, especially if a companion is tethered near, or they are within sight or sound of humans.

A fairly wide collar, fitted so that she cannot pull out of it, will be adequate. Leather is very suitable and remains pliable as it is continually greased by the goat's natural oils. A greyhound collar, obtainable from pet shops, is about the right size for most goats (see Fig. 38).

By far the most common length of tether used is around 12 feet. This will ensure that the goat is moved frequently enough to allow her to feed at the

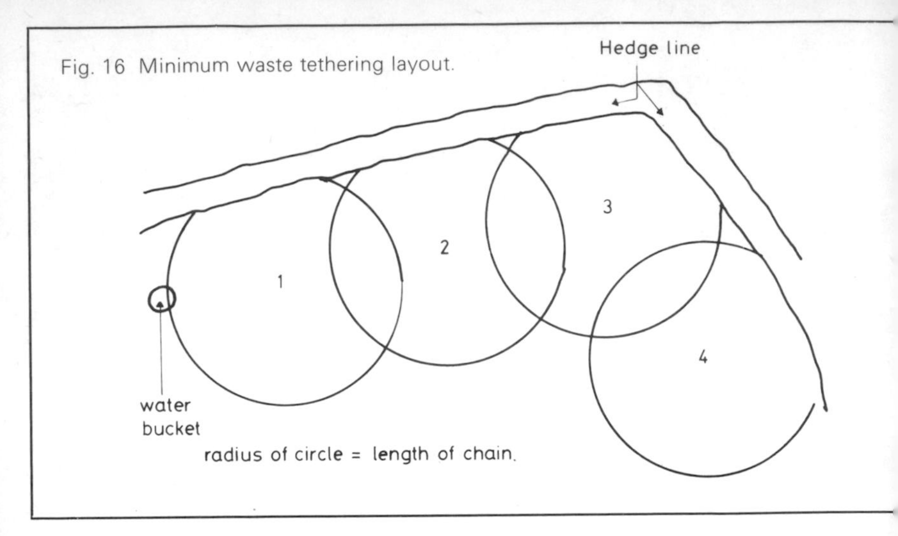

Fig. 16 Minimum waste tethering layout.

end of her tether (as she inevitably will) and not damage too much underfoot. It is not necessary to move a whole circle at a time. When selecting a spot to tether the goat ensure that she cannot become entwined in any hedges or stumps. Pace out the distance and tether accordingly. If she bleats when staked out this may mean she does not approve of the spot you have chosen for her. For a contented goat *and* goat keeper try her in a different spot. She will soon indicate that all is well by tucking into the browsings available if the area is up to her expectations. Figure 16 shows a suitable tethering layout.

Chain is more suitable than rope as it will stand up to the wear involved and is less likely to knot. A chain eighteen links to the foot, eight gauge is adequate. A swivel at either end will avoid the chain knotting. (See Fig. 17.)

A running tether may be more economical in grazing flat land. The wire may be any length from a few feet to a hundred or so yards. The stakes at either end are driven into the ground holding the wire taut. The goat is attached to a short stout chain (six links to the foot) one foot long with swivel ends and clipped onto the fixed wire. A short stop stake to prevent the goat running the entire length of the wire can be moved as required.

The goat will be happier moved several times a day under both systems of tethering. Do not forget to allow her fresh clean water at all times while she is out. Arrange to leave the bucket at the end of the tether to prevent it being knocked over prematurely. We use discarded motor tyres in which to stand the buckets. They are easily transported and keep the bucket upright under most conditions.

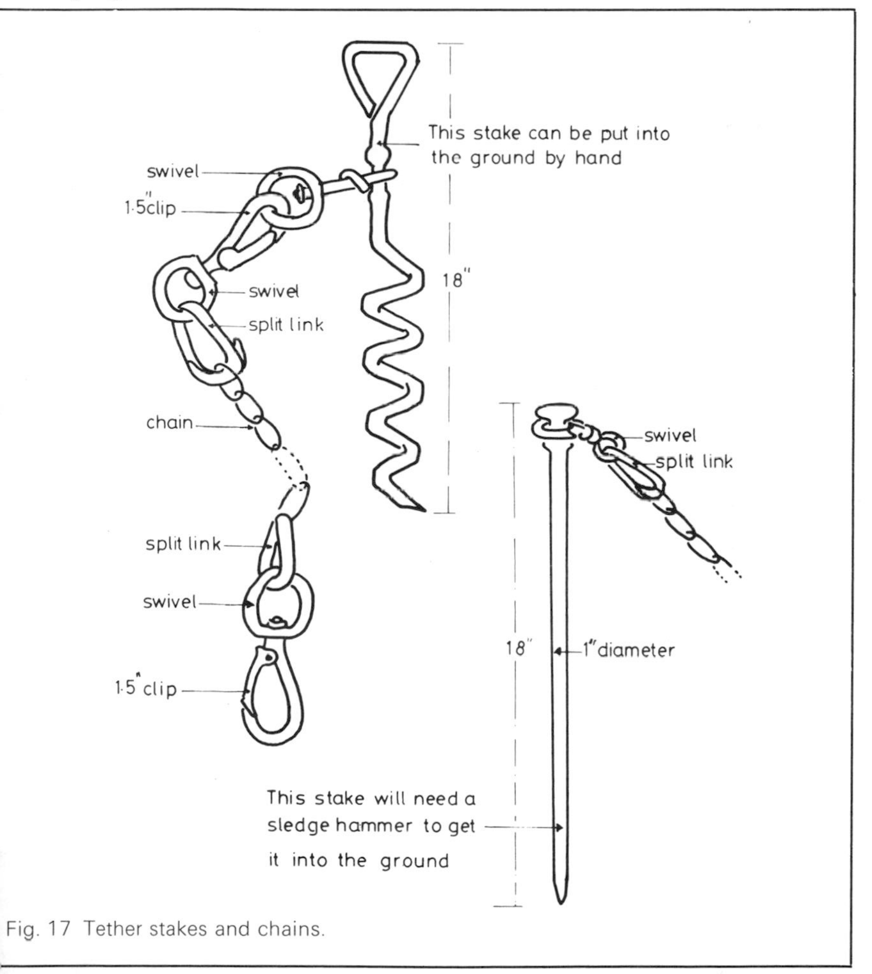

Fig. 17 Tether stakes and chains.

3 Food and Feeding

Each goat is a law unto herself as far as food is concerned. Rations which would be palatable to one may be refused by another, though this choosiness is more common among the higher yielding goats. On the other hand, foods which may have no effect on the bowels of the average milker might result in scouring in the highest yielders. The observant goat keeper is therefore at an advantage.

Goats are highly conservative and may go hungry rather than eat a particular food to which they have been newly introduced. Watch her coming to terms with a new food for the first time. With legs apart, ears well pricked, she looks inquisitively at the food, body poised for flight. A snort, a nibble, a spit (and goats are second only to camels in their ability to show disapproval in this way), a nibble and a chew. If it comes up to expectations, she may swallow. After a few days, a new food which has passed the test will be consumed with avidity.

The attitude to her feed bucket also merits attention. Some goats would no more dream of eating from a loose metal bucket on the floor than fly. Others may eat out of anything, anywhere.

Goats being ruminants spend alternate periods browsing and chewing the cud. A certain amount of time is spent eating a mixture of both herbaceous and woody plants. This is followed by a more settled period when the animal either lies down or stands quietly chewing and regurgitating alternately. The animal then starts to feed once more. The goat is designed to eat vast quantities of fibrous foods. She crops and swallows the food; this coarsely masticated material is stored in the rumen and soaked by liquid from the second stomach till she is ready to cud.

When cudding starts a pellet of food is regurgitated to the mouth, and is chewed up to seventy or so times before being returned to the rumen once more. The more fibrous foods may be regurgitated several times before passing eventually to the fourth stomach where digestion takes place as in the stomachs of simple-stomached animals.

A goat has twice the rumen capacity of a sheep, and provided we can keep her supplied with enough roughage foods, she will remain healthy and productive. Her large appetite for both herbaceous and woody plants, which are higher in mineral, protein and fibre content than grasses, makes sure that her needs for milk production, as well as maintenance and growth, are met. The weeds and scrub that other farm animals generally spurn are utilised and, in fact, make a valuable contribution to her health. A ration consisting of a high proportion of grain foods and good quality clover and rye grass hay, which would satisfy a high yielding cow, can produce an unthrifty goat predisposed to chills and disease; the animal would be extremely expensive to feed into the bargain.

Water is essential to life. The goat must consume a sufficient quantity daily from the water bucket besides the water that is in the succulent foods. Water is one of the components of many essential body secretions; nutrients are carried round the body in the bloodstream which consists mainly of water. Water is passed out when the animal breathes and urinates. Milk also contains 86 per cent of water. The body temperature control is largely adjusted by evaporation of water from the lung surfaces and, to a lesser extent, through the skin, so don't be surprised if, in hot weather, she pants rather like a dog. Therefore, she *must* have access to adequate, clean water at all times. If she does not drink as much as is expected, a smell, undetectable by a human being, may be putting her off. The water container should be changed for another. Many animals prefer plastic rather than metal, on the other hand it may be the other way round! Scrub out the water container daily, and change the water frequently. A goat fed a high concentrates ration likes her water warmed in winter or she may suffer from the cold. This is due to her reduced roughage intake and consequent low level of bacteria, a by-product of whose activity produces considerable heat in the roughage-fed goat.

Foods may be divided into two main groups:

1. Concentrates
2. Bulky foods or roughage

The domestic goat has been bred for possibly thousands of years to produce more milk than the needs of her kids. In spite of her vast capacity for bulky foods one must augment this with concentrates to provide for her very considerable energy needs. A high standard of management will be required if much concentrates are fed. Management is simplified if 2 pounds of concentrates are fed daily together with a wide range of bulky roughage foods fed to capacity. In summer this will mean as much browsing and grazing as she can manage and in winter up to 5 pounds of mixed hay (plenty of nettles and hedge weeds included) as well as succulents such as cabbages.

Many mills will supply concentrates in the form of a coarse dairy ration for goats (a finely-ground meal forms a pap in the mouth and is unsuitable). This can consist of rolled oats, crushed barley, flaked maize, bran and other foods such as linseed cake, soya bean meal and locust bean pods mixed to their own formula and correctly balanced for milk production. Alternatively, a dairy nut intended for cows can be bought.

This is a ration balanced for milk production which is sold in the form of pellets, known in the trade as dairy nuts. Goats do not find this form of food very palatable, perhaps the nuts are too large. Calf weaner nuts are made in a smaller form which goats will consume with relish and which are also higher in protein content.

Some goat keepers like to mix their own goat ration. It is worth experimenting to find out what the goat really *does* like. This is best done gradually and preferably not at the beginning of the lactation when changes in the diet can have a disastrous effect on the milk yield. The greater the variety of constituents in the ration, the less likely the animal is to suffer from deficiencies.

Concentrate foods are expensive, especially the proprietory mixes. These proprietary mixes have added vitamins which the maker does not guarantee will be operative after a certain period of storage, and this date may be marked on the sack in which the food is packed. This means that when buying foods, you must make sure that they are finished, usually, within a few weeks, as any grain will begin to deteriorate after it has been rolled or ground. It may be convenient to share a sack with a friend who also keeps goats. Buying food by the ½ cwt is cheaper than smaller quantities (especially if it is collected from the mill and cash paid for it on collection). Buying from a pet shop from open bins gives no guarantee as to how old the mix is. The advantage, however, is that smaller quantities can be bought. Generally, millers prefer to sell in ½ cwt lots or more.

Freshly ground foods are always more palatable. I like to buy concentrates at least every ten days in summer, and twice monthly in winter when the food may not deteriorate so quickly. A cool, dry, securely lidded storage bin will enormously assist the keeping qualities of the ration and will prevent the depredations of both rats and mice.

Table 1 Sample food costs per ton with equivalent costs per pound

£/ton	pence/lb
10	0.45
20	0.9
30	1.35
40	1.8
50	2.25
60	2.7
70	3.15
80	3.6
90	4.05

The concentrate ration will need to be fed at least twice a day, and in a high yielding goat three or four times a day may be advantageous. It is convenient to feed some of it while the goat is being milked, this will help ensure her co-operation and she will be more likely to stand still.

The bulky foods should form the main part of the goat ration. They are of two main sorts: succulent and fibrous. These, with a little resource, can be obtained extremely cheaply or even free, they are also vital to the goat's well-being and health.

The grazing goat will tend to select such plants as creeping thistle, hogweed (cow mumble), plantain and dandelion. These plants have a higher protein content than grass and their various depths of root penetration can make minerals available from a great range of soil levels. Chicory and dandelion, for instance, having long tap roots, bring minerals up from some distance below the surface, while surface-rooting plants, like nettles and groundsel, make the minerals from just below the surface available to the goats. A goat put out to graze on a good dairy pasture would be very unhappy without these weeds to

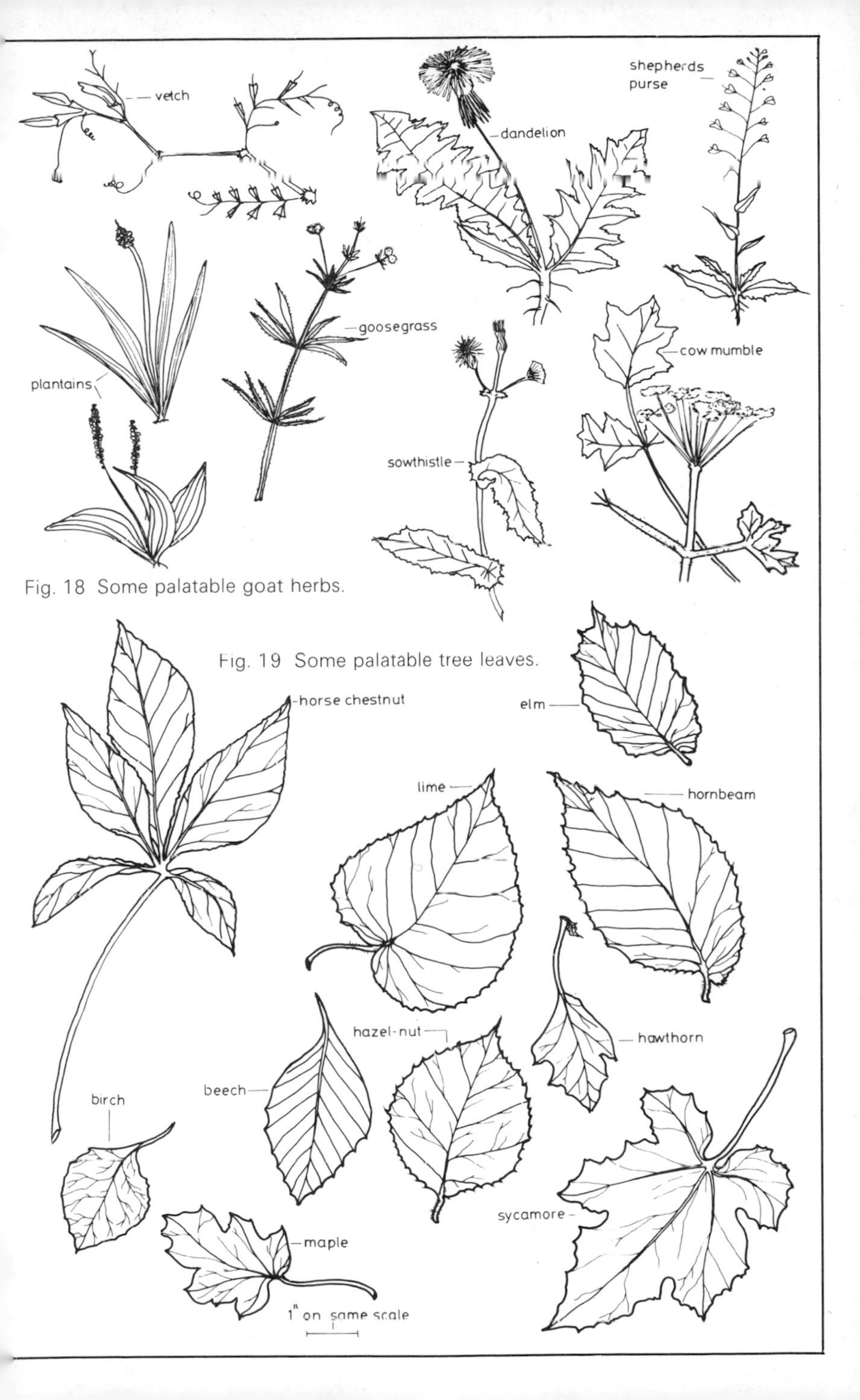

Fig. 18 Some palatable goat herbs.

Fig. 19 Some palatable tree leaves.

supplement the grass. Figure 18 shows some palatable plants.

Browsings can form a large proportion of the goat's roughage diet and include, for example, the leaves of elm, lime, ash, beech, quickthorn and apple, which are particularly palatable, also alder, willow, cherry, maple, sycamore, birch, plane, hazel-nut and sloe. Oak leaves offered in very small quantities will soon bind up an animal which is scouring slightly, whereas leaves of the ash tree will loosen the bowels of a constipated animal. Bramble, gorse and holly are also eaten with relish in spite of their prickles, also broom and heather. Some palatable tree leaves are shown in Fig. 19.

For the goat which is unable to wander at will to find all these leaves, the materials can be cut and brought back for her to eat. Judicious pruning of trees can improve their shape as well as providing goat roughage. Care should be taken not to prune trees excessively, and they should be yours anyway!

Branches can be tied to a hook in the goat shed to keep them off the ground (see Fig. 20). A kid stool placed below the branches will help the smaller goats reach them. I like to gather a mixed selection of browsings to ensure that a wide spectrum of minerals is available.

Nettles are a particularly nutritious food, being high in protein and iron. The goat will nibble the mature flowers and as the autumn progresses will eat the plant down to the ground. I like to cut the nettles down as they come into flower. This is done when it looks as though there will be a few dry days when the sun will dry the cut plants. When the plants are dry and the leaves are blackened they are ready to harvest. In the evening, just as the dew begins to rise and the leaves are slightly damp, I gather them carefully, avoiding losing too much leaf through rough hand-

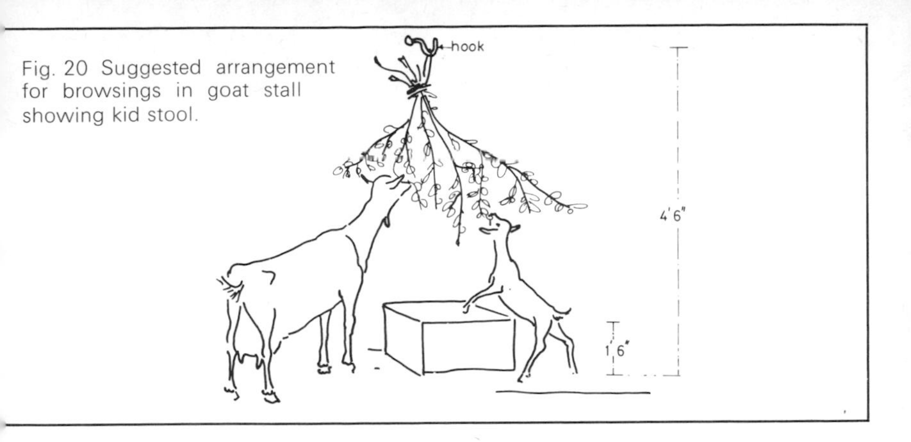

Fig. 20 Suggested arrangement for browsings in goat stall showing kid stool.

ling. They are then stored under cover for the winter, laid out on opened paper meal-bags to catch the inevitable broken leaves which fall.

The plants that grow in our hedgerows and road verges make a superb goat hay. Containing vetches, nettles, hogweed and other nutritious plants, it forms a valuable addition to the winter diet if cut and dried carefully as the plants come into flower. Turn daily, or more frequently if possible, and cart and store it when the hay is dry. A wooden floor, or failing this, a thick layer of straw or hedge trimmings will make a good stack bottom for the hay. Avoid storing hay directly on concrete or damp earth.

During the spring and summer, do pay attention to the crops which may be being sprayed in the fields adjoining the hedgerows. A tactful word with the farmer and he will tell you when he is going to spray and if the spray he is using will affect any adjoining grass verges. Remember also that busy road verges may be not only directly dangerous to harvest because of heavy traffic, but also may carry toxic levels of lead deposit on the herbage.

Friends with a weed patch in their gardens are often only too pleased for the rough grasses and weeds to be cut, made into hay and carted away. This is best done in early June when the plants still have a high protein content. Baled rye grass and clover hay which would delight the dairy farmer would not produce the same response from the goat keeper. The goat likes coarse grasses, in fact Yorkshire Fog, a grass which is abhorred in hay made by farmers, is highly palatable to the goat.

During the summer, weeds found on the roadside verges, and browsings, form an adequate diet in sufficient quantity. This is fed in conjunction with the basic 2 pounds of concentrates. But as autumn draws on, the browsings are reduced and the succulents are not so freely available. The goat will change her feeding habits and the milk production will drop over a couple of weeks. The goat will become accustomed to filling up on much drier roughage, and hay will begin to play a larger part in her diet. As she gets accustomed to the winter feeding programme and the weather gets colder, she *may* lay on more fat. The milk yield will then level off once more.

In winter, bramble leaves, holly, ivy (but *not* the berries) will be obtainable to supplement fresh foods for the goat. Bark of younger trees will be very popular with her but this may not be appreciated by the keen gardener. In winter a cut-over woodland will provide roughage in the form of saplings, brambles and weeds for one goat per acre, but mature woodland will yield far less. This winter roughage can be supplemented by the weed and coarse grass hay made the previous June.

Kale, mangolds, turnips and fodder beet are palatable but need to be fed with plenty of hay as the water content is very high in these succulents. Roots such as these can be fed up to 6 pounds a day, but remember to avoid the mangolds and fodder beet until after Christmas. Sugar beet is enjoyed by goats but under no circumstances must the tops be fed until they have been *thoroughly* wilted. Turnips and kale must be fed *after* milking or the milk will be strongly tainted. Any of these foods should be introduced gradually or scours will rapidly develop. Potatoes boiled, mashed and mixed with bran are enjoyed by goats in the winter, while brussel sprout and cabbage stalks split down the centre will also be relished. When feeding brassicas from the garden, make sure to burn the remaining root if the plants are suffering from clubroot.

Dried sugar beet pulp can be bought by the 8 stone bag and this can be fed either soaked or dry. It will keep all winter provided the sack is stored dry on a wooden board (it goes mouldy if stored on concrete). It is high in calcium, but being low in phosphorous is best supplemented with bran. No more than 2 pounds a day should be fed.

The goat with plenty of roughage in her will keep warm and fit, but she must not have too warm and stuffy a shed if she is to go out of doors during the colder weather. The stall-fed animal needs to be kept warmer as she is unable to generate heat by moving around. A high concentrate diet will supply the energy needs for this purpose.

As the days begin to lengthen, improve the quantity, quality and palatability of the ration, or, if she is in kid both the goat's milk production and her subsequent lactation after kidding will suffer. Her milk production will naturally improve with the lengthening days. Without improving the ration she will milk to the detriment of her body condition. If the goat is to be dried off prior to kidding, cut down on the succulent feed—roots and brassicas—and make her ration as dry as possible (see page 54).

The more naturally the goat is fed—browsing and grazing on rough land—the less likely she is to suffer from mineral deficiencies. Producing such a vast quantity of milk in relation to her size, any mineral deficiency quickly makes itself apparent. The plants and trees which the animal finds most palatable are naturally high in the minerals she requires.

It is essential that the stall-fed goat has access to a mineral lick. There are several of these on the market, ranging from the natural rock salt which supplies sodium, an iodized salt which has iodine added, to more comprehensive mineral blocks using salt as the base. A proprietory dairy ration formulated for goats will generally contain some added minerals.

The goat fed on a more natural ration containing adequate natural roughage both in winter and summer will rarely suffer from mineral deficiencies, but access to a good mineral block is still advisable as an insurance. Where soil is very sandy and both rainfall and drainage excessive, mineral deficiencies may arise. The value of a wide-spectrum mineral block being available at all times cannot be overemphasised.

Some mineral blocks are sold with a metal container which can be tacked onto the wall of the goat shed; others have a hole through them to enable them to be suspended from a hook by means of rope (see Fig. 21 for examples). A mineral block which is allowed to be fouled by the goat in any way will not be licked, however badly she needs the minerals.

Cobalt is particularly necessary and readily available in chicory, plantain, sorrel and heather. A deficiency shows itself in poor tasting milk. This can be rectified by dissolving 1 ounce of cobalt sulphate in ½ pint of water. This is poured over a container filled with 6 pounds of common salt which is then allowed to dry. This can be offered daily to the goat and she will consume it avidly if her diet is deficient in this mineral.

Another mineral commonly lacking is iodine. The animal develops a dry, starey coat; the skin, particularly that of the udder, will resemble parchment.

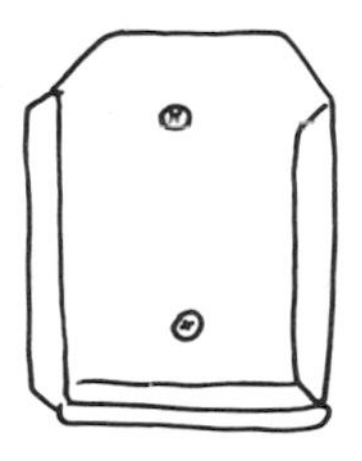

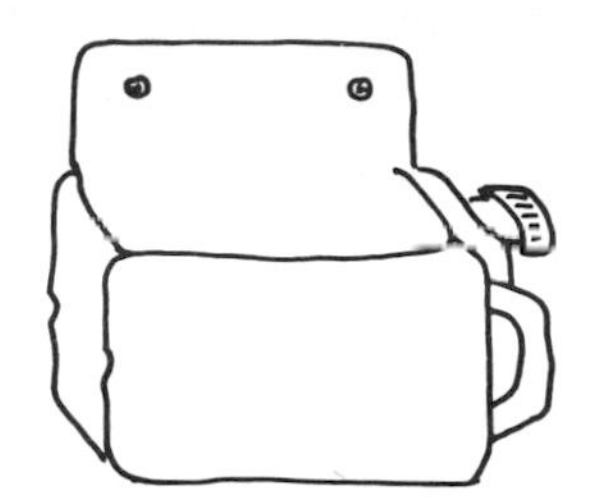

Fig. 21 Mineral block holders.

Another frightening development may be the production of dead female kids while the male kids thrive. It also affects the animal's ability to assimilate Vitamin A and this reduces resistance to infection generally, growth is slowed and infertility can result.

It cannot be emphasised too strongly that the diet rich in *adequate* natural roughage, in the form of weeds and browsings which provide sufficient minerals and vitamins, will vastly reduce the chance of any vitamin and mineral deficiencies.

A free-ranging goat is less liable to poisoning than the stall-fed or tethered goat. The free-range goat in the herd follows the feeding example of the leader who in turn has been taught by his leader. It is best, of course, to prevent access to poisonous plants by not tethering near to the offenders or even destroying the plant where possible or convenient. Unfortunately privet has a fatal fascination (*literally* for goats) especially for those that are tethered. They will move heaven and earth to reach it when passing. The entirely free-range goat is apparently more selective and will pass by this plant with a sniff (see list of other poisonous plants page 97).

For those of us with spare land in the garden, there are some crops that can be grown specifically to augment the goat's diet. Kale, especially the variety Maris Kestrel, which retains its leaf even in the hardest winter, is palatable down to the root, and will keep until April; it is well worth growing. Sow preferably before the Grand National race (around March 12th) or in early June. Drill it in rows 18 inches apart in land well mucked the previous autumn. Keep it well hoed till the plants have met in the rows. Thin the plants to 6–12 inches apart.

Fodder beet (variety Red Otoffte) is a particularly useful root though it must not be fed till after Christmas. Sow a little later than the kale in drills 18 inches apart and thin plants to 6–12 inches. A 20 yard row of well-grown fodder beet would supply about 2 cwt of roots; this would be adequate for two goats in January and February.

In land free of perennial weeds Russian Comfrey (since it is not an aggressive competitor) is a good green succulent for feeding from April to October. Set the plants 3 feet apart in well manured land. After a couple of years these plants will yield up to five cuts of fodder per summer if they are not allowed to run to flower.

Pea and bean haulms, the bi-products of household vegetable production, are greatly appreciated by goats, as well as fruit-tree prunings, rose prunings and hedge trimmings (which do not contain any of the listed poisons on page 97). Household vegetable parings, possibly boiled if potato skins are included, can all be used. Potato peelings, baked dry in a cool oven and mixed with the concentrate ration, are considered a great treat by some goats.

Lawn mowings can be a useful addition to the diet, provided that they are free from poisonous weeds. They must be sun-dried or *absolutely fresh.* Mown grass which has been allowed to heat in any way at all must *not* be fed to the goat—this will severely upset the goat and could possibly kill her.

Some foods may taint the milk, this especially

applies to the brassica family which includes turnips. Sugar beet pulp may impart a slightly fishy flavour to the milk. An animal fed a natural high-roughage diet may develop characteristic flavoured milk. One soon learns which plants and trees affect the flavour. Elm has little apparent influence on this, but lime leaves fed just before milking may make the milk taste slightly bitter. Sheep's parsley, yarrow and charlock can also taint the milk, so access to these plants is best confined to just after milking time. Goats respond individually to different plants, some may produce a sweet flavoured milk whatever they are fed, but others fed exactly the same diet may produce 'goaty' flavoured milk. Oats and carrots, on the other hand, are said to improve the flavour of milk.

Most goat keepers will be unable to keep a goat on free range entirely, so a combined stall and semi-free routine will be the most suitable. This will involve some planning as to how it can conveniently fit into the existing household routine. Bearing in mind that feeding little and often is the motto, the following might be a suggested daily schedule:

7.00 a.m.	Hay and fresh water.
7.30 a.m.	Milk. 1 lb concentrate ration. Browsings in summer. Kale, brassicas in winter.
10.00 a.m.	Exercise in yard, on tether, or a walk round grass verges on a lead.
12.00 noon	Hay and water.
4.00 p.m.	Hay and water (no kale or taint making foods).
7.00 p.m.	Milk. 1 lb concentrates followed by cut browsings in summer, brassicas in winter *after milking*.
10.00 p.m.	Final check. Top up water. Top up hay.

A goat attended less often will not yield so well, but will survive and may still be profitable. An animal tended but twice a day will not give of her best. It is difficult to generalise on the quantities of hay, browsings, and kale needed. The aim is to supply the goat with enough for her to have finished before she is fed again. If any is left, she has either been offered too much or she does not like the particular food.

Feeding guidelines:

1. Try to be as regular as possible. Animals have a built-in 'clock' and adjust their habits to a fixed routine.
2. Introduce new foods gradually over a period.
3. Clean food and water containers regularly.
4. Clear away any food left from previous feed.
5. Avoid feeding frosted foods.
6. Watch the animal feeding—a goat slow to eat may be going downhill.
7. Observe the smell of her shed on entering. A normal animal smell will soon be recognised; a sick animal develops an unpleasant, possibly sour odour.
8. Observe droppings for any change in consistency.
9. Feed tainting foods directly after milking.
10. Pay particular attention to the goat's particular likes and dislikes as far as different foods are concerned.

A semi-stall-fed goat requiring hay all the year round will require about ½ ton of hay. It is difficult to say how many bales this is, as they vary enormously in weight. On average, twenty-five bales or a lump of home saved, unbaled hay 10 feet square and at least 6 feet high (when fully settled) would see one goat through till the next hay harvest. It is worth buying plenty of barley straw for litter as this can always be used for feed if hay runs short.

It is difficult to generalise on amounts of food needed, but allowing for the daily 2 pounds of concentrates, and hay, the other foods will be fed to appetite paying particular attention to the individual's likes and dislikes and the fact that little and often will achieve the best results.

4 Breeding and Milk Production

In temperate countries the goat, like the deer, confines its breeding season to the autumn. The billy, who has a strong goaty smell at the best of times, becomes even stronger scented. The female comes into season (or 'heat') approximately every three weeks from about August (although the signs are not easily noticed early in the season by the inexperienced goat keeper) till December and January.

In Britain we generally wait till the goatling is at least eighteen months old before she is taken to the billy. Unless she is extremely well grown, an earlier mating will result in stunted growth in the mother, small kids and less good milk production.

The gestation period is approximately 150 days, a week or ten days early or late not being unusual. If the goat is mated early in the autumn, the kids will be born correspondingly early in the next year when there is less succulent feed about. This situation will mean that drying off the goat for a six to eight week period before kidding will be simplified as she will be accustomed to a largely dry and fibrous ration (see page 44).

From August onwards, watch the goat closely for signs of season. Having decided when you would prefer the goat to kid, you can then calculate roughly in which month she will need to be mated (see Table 2). If the first date of heat is noted, the next time can readily be calculated by adding on three weeks to the first apparent heat. Most animals come into heat fairly regularly.

Table 2 Weekly gestation table

Mating date		Approximate kidding date		Mating date		Approximate kidding date	
July	1	November	28	November	4	April	3
	8	December	5		11		10
	15		12		18		17
	22		19		25		24
	29		28	December	2	May	1
August	5	January	2	December	9	May	8
	12		9		16		15
	19		16		23		22
	26		23		30		29
September	2	January	30	January	1	May	31
	9	February	6		8	June	7
	16		13		15		14
	23		20		22		21
	30		27		29		28
October	7	March	6	February	5	July	5
	14		13		12		12
	21		20		19		19
	28		27		26		26

If you are well acquainted with your goat's general behaviour, it will be all the easier to notice any change. Probably the first obvious signs will be more frequent bleating than usual. If you know that she has all the food, water and general attention she needs, then look for further signs. She will wag her tail more frequently, her vulva will appear a little swollen and she may even be discharging a little clear liquid. If already milking, she may suffer a slight drop in milk production, though this is not always the case. Her general attitude will be restless. Some animals show only slight signs and these for a very short period (just a few hours), while others are very obviously on heat and remain so for at least a day. If there is any male goat living near, do not leave the goat out of doors unattended, she might break loose and find her way to him, or he may even pay her a visit, resulting in a possibly unwanted pregnancy.

Each year the British Goat Society issues a Stud Goat Register of pedigree male goats available for general use. It lists males which are known to sire female kids producing more milk and butter-fat than their dams. On the other hand, if the pregnancy is only required to renew the milk supply and the kids are to be destroyed at birth, one will only have to look for a billy of known health and fertility.

Having decided on the billy to which you wish to take the goat, having already made arrangements for her to be mated and warned the owner of the time of your goat's expected heat, contact the owners and tell them the time you expect to arrive. A car is a convenient way to take her, but if it is no further than a mile she can easily walk if not hurried too much. The road surface will do her feet good! She can even travel in a trailer behind a cycle if accustomed to this mode of transport!

The sooner you get to the billy the better! The billy will mate her if she is still on heat. It is normal practice to pay for the service before returning home. Before leaving ask for a certificate of service stating the name and number of both nanny and billy. This will be needed for the eventual registration of any resulting kids.

Make a note on the calendar in the meal store or milking shed when she will be due to come on heat again. If she has been successfully mated and shows no signs of heat three weeks after mating, you may be fairly sure she is in kid. If she appears to be on heat again a repeat service will be required.

Care must be taken of the in-kid goat. Don't hurry her unduly, especially when passing through doorways when she may get knocked. Keep to her usual daily routine. She should have been treated for worms before mating, but if you do suspect a worm infestation she can be treated during pregnancy as long as she is handled gently. See that her feet are well trimmed before she begins to obviously put on more weight as the kids develop. She relies on sound feet, even if indirectly, for so many essential activities.

Pregnancy is a perfectly natural process and in the vast majority of cases all will go well. A normal feeding routine is continued, paying particular attention to providing as wide a variety of food as possible. If the animal is already milking she may show a decrease in yield when she has been mated.

Exercise is important to the in-kid goat especially one that is stall fed. The temptation is to decide that it is too cold, wet or generally wintery for her to venture out. Obviously the tethered animal (with access to a shed) gets more exercise than the stall fed animal, but they will both need enough to keep them healthy. The goat which lies about in her shed is more likely to end up with malpresented kids which may result in a difficult kidding.

I like the goat to be dry for at least six weeks before kidding to allow her to build up her resources for the strain of kidding and the subsequent lactation. Drying a goat off in winter is easier than in spring as her diet will already contain less succulents and a higher proportion of hay. Restriction of water by only offering water before feeding twice a day, dry high fibre diet, and cutting out all succulents will help. Incomplete milking, and possibly milking only once a day for three days or so before leaving her for a couple of days with no milking, can also be effective. A method I find which works well involves someone other than the normal handler feeding and milking the animal. The sight of the regular milker at the appropriate time is enough to stimulate production once more. I ask my husband to feed the animal while I keep well out of hearing and sight of the goat.

After a few days the full udder will begin to slacken at the top and only look plump near the teats (see Fig. 22). This is a sign that drying off is proceeding well. Once more you may visit your favourite milker. *Never* leave the goat uncomfortably full with the skin tight, red and hot. A few squeezes to ease the udder will be sufficient to relieve the situation. If the udder is allowed to get really hard mastitis may set in (see page 104). Goats from high yielding strains are very difficult to dry and some may never dry completely before kidding.

As drying-off proceeds, the increased demands on the goat of the rapidly developing kids must be considered. Gradually increase the concentrates, which will have been withheld during the drying-off period, to 2 pounds a day at a month before kidding. Too much cereal will result in kids too large for comfortable kidding. Maintain a plentiful variety of bulky foods.

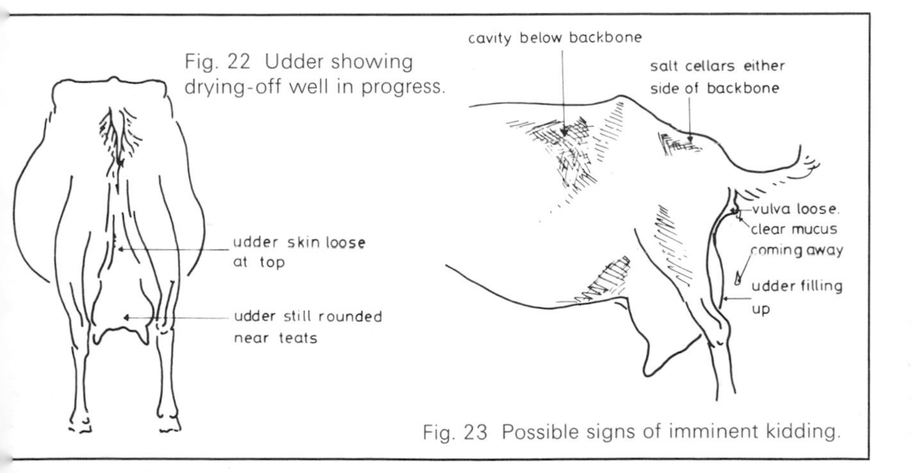

Fig. 22 Udder showing drying-off well in progress.

Fig. 23 Possible signs of imminent kidding.

As the animal gets nearer to kidding increase the succulents. This will coincide with more being available as spring advances and the days lengthen. Marrow-stem kale, old cabbages, roots, mangolds and wet sugar beet pulp will fit the bill here.

A goatling showing very little udder may rapidly develop any time in the last six weeks of pregnancy. She may become extremely uncomfortable, and in this case she will need to have some of the milk eased off to relieve the tension till the skin feels less tight. While the udder is developing rapidly you may need to ease a little liquid from her every other day, especially if she is from a high yielding strain. Don't worry if one side of the udder is larger than the other. This is perfectly normal and will right itself within days of kidding. When milking one side because of the fullness, very gently milk the less developed side even if no milk is being secreted. The manual stimulation will encourage it to develop and will help to balance her up. Great care must be taken when doing this to prevent the organ being damaged.

Some goats will not need pre-milking at all as they develop more slowly. Pre-milking in no way affects the production of the antibodies in the colostrum necessary for the well-being and health of the newly-born kid. However, milking the goat up to the time of kidding will possibly adversely affect the quality and quantity of the subsequent lactation in the case of the goat who has already had kids.

A few days before kidding, the vulva of the goat will loosen and soften in conjunction with the slackening pelvic ligaments (producing salt cellars) on either side of the tail (see Fig. 23). When the goat is lying down, the vulva will protrude considerably. The kids will increase in activity up to twelve hours or so before birth, then the goat will develop hollows in her sides and hardly any movement will be felt. The sides of the goat will feel much firmer than usual. She may appear restless and will not be eating so well.

A few days before the expected date of kidding cut out or considerably reduce the concentrate ration which can stimulate too much milk too soon after birth. This can shock the animal's system too much and possibly cause milk fever in the older goat. By increasing the succulents and fibrous foods, she is not so likely to become constipated. The addition of a few ash leaves, if available, to her diet will ease the situation if she does have a little difficulty with bowel movement.

The box for the kidding goat needs to be clean, light and airy. Some people prefer to have a special loose box ready for this with no internal fittings which might get in the way. On the other hand the goat's own living stall, if it is reasonably roomy, will be adequate if it is cleanly and deeply littered and is free from protruding fittings.

I like to remove the water bucket last thing at night near kidding time, in case it should result in an accidental drowning of the new-born kid. Coarse peat moss is an ideal litter or reasonably deep soft straw (barley straw is good here).

During the last few days before kidding she will progress in a series of stop-go stages. At one moment it could be imagined that she will kid imminently, and then everything seems to stop again. She may well wait for a fine, warm, sunny day. During the last few hours she will be loth to exercise and her feeding will be fussy and her cudding fitful.

A well-handled goat with an equable temperament may want her owner to be present at the kidding. As signs of imminent birth are noticed and she is seen to strain and then continues to cud, it is a good plan to time the gaps between the straining. If the strainings (contractions) come every three minutes or so, birth may be very near. She may bleat alarmingly with each contraction, but some goats are quite silent at this stage. Her vulva will be extremely loose. Her udder will feel hard and when she tries to defaecate she will raise her back unusually. With contractions every three minutes, her breathing will be shallower. She may walk about and lie down alternately, while her ears may hang sideways and down in an unusual manner (in breeds other than the Anglo-Nubian). Her eyes will be a little more wide open than usual.

The clear discharge, which may have started intermittently a week before, will possibly increase. The water bag will show out of the vulva like a creamy coloured balloon and this is when she may want comforting. If, on the other hand, she tenses each time she is looked at, leave her alone but still continue to look in at her unobtrusively at intervals.

Some goats give birth quite happily in a standing position, while others prefer to do so lying down. The goat which kids in a recumbent position will push her back legs out to one side in conjunction with each contraction. A normal birth may then be expected quite imminently. The salt cellars, noticed earlier on either side of her tail, will be filled as the first kid enters the vagina. Some goats will cud quite happily during the whole of this stage except while straining. A few pushes will expose the two front hooves and the nose first, followed by one or more stronger contractions (depending on the size of the kid relative to the birth canal) which will propel the kid into the world. Figure 24 shows normal presentation of a kid.

The vigorous kid encased in its water bag will break out, cough and shake its head; the breathing will be shallow for a few minutes after birth and it may start to bleat. The goatling, however, may not show any interest in the new-born kid until this bleat is heard. If not already standing, she will get up and look around,

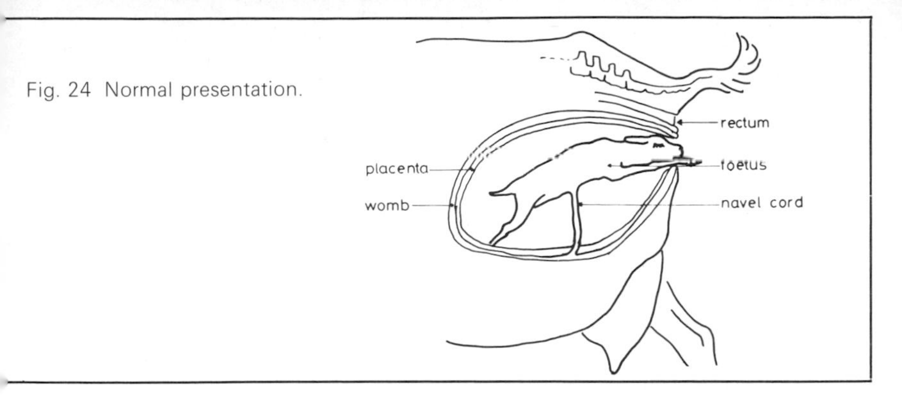

Fig. 24 Normal presentation.

and this will sever the umbilical cord. She will sniff the creature and begin to thoroughly lick it all over. If another kid is due it will probably be born within half an hour or so of the first. If the goat shows no interest in the kid even after hearing it bleat, the animal will have to be dried by hand to remove the mucus and stimulate the blood supply. It will need to be fed the vital colostrum from a pre-heated, sterile bottle as soon as possible.

Some kids are exhausted by the actual birth process and show little inclination to breathe or even move. In this case gently remove the mucus from its nose, mouth and even its throat with a clean finger. If it still shows no sign of breathing, hold it up by the back legs and give it a *gentle* shake. This will usually shock the creature into normal respiration.

After the last kid has been born the mother's behaviour will become more normal, she will lick the kids energetically and this often stimulates them to attempt to stand. Some vigorous kids will stand within minutes, whereas the more lethargic animal, possibly from an older nanny, may not stand for half an hour or so. I like to lead the kids to the udder while I squeeze a little milk onto their noses. They will soon lick their lips and then begin to seek the teats. A goat with large, fat teats may need to have some of the milk eased off, in order that the kids can grasp the teat comfortably.

As soon as the kids are born, some people like to put tincture of iodine on the navel which may help to dry it, and it will also have some disinfectant value. Check the sex of the kids and in the case of the females make sure that the teats are normal. Supernumerary blind ones can be cut off with surgical scissors and the scar disinfected. If a nanny kid has double or additional normal teats it will have to be destroyed. However, the condition is not at all usual.

If a billy kid is not wanted for meat at a later date, it is kindest to kill it as soon as possible, but do let the mother lick the kid as this may help the contraction of her womb and the subsequent release of the placenta (or 'afterbirth'). To kill the billy kid, put a few drops of chloroform on a piece of cotton wool just inside the mouth of a kilner bottling jar. Hold the kid near the cotton wool and as it loses consciousness push the head right into the jar. Place kid and jar in a box or plastic fertilizer bag and close it for half an hour to ensure that the animal is absolutely dead. Such a job is best done away from the goat and outdoors to avoid the risk of inhaling chloroform fumes yourself.

The afterbirth is expelled from the vulva, as the name implies, soon after the kids are born. In a normal birth it will look reasonably healthy and untorn. In a difficult birth, part of the placenta may be retained. There will be one for each kid unless they are identical twins when the placenta will be shared. It will normally be produced up to twelve hours after the birth, if any later the vet should be called to investigate. Some like to remove the placenta after it is expelled, but I prefer to leave it for the goat to eat if she likes. This is quite natural and also helps the womb to return to its normal size.

It is essential that the kids suck the teats within six hours of birth while the antibodies in the colostrum are present. This will protect the newly-born animal.

After parturition the milk is richer in solids and thicker in consistency than normal milk. This is known as colostrum, or regionally as 'beestings', 'beastlyns', 'firstlings' or 'strokelings'. It contains a high proportion of protein which is especially suited for easy assimilation by the kid. It is also a laxative and contains vital disease immunizing factors which are passed from parent to young. If boiled it will curdle so if it needs to be given to a bottle-fed kid, it is wisest to milk directly into a sterile, pre-heated bottle to prevent it cooling and to ensure that the kid receives the milk at the correct temperature. If it does have to be heated, it is easiest done in a double saucepan. Five days after kidding the milk is considered normal.

Most kiddings proceed naturally with no complications. From one to five kids may be born, the usual number being two or three. Malpresentations are best handled by a veterinary surgeon. When an animal has been straining strongly for an hour with no visible signs of a kid, or perhaps the feet have been showing for at least half an hour with no further progress, expert help is needed. Manual assistance in kidding may result in infection and this will result in the animal developing a temperature. Your vet will be needed to treat this.

After kidding offer the goat some warm water as she may be extremely thirsty. Some goats will like a bran mash while others will prefer a little good hay. To make a bran mash, sufficient boiling water is poured over a double handful of bran in a bucket to make a crumbly mix. This is covered with a cloth, left to cool and then offered to the goat while it is *just warm*. The mash will help to assuage her appetite and will have a laxative effect. Do not feed a concentrate ration for at least the first two days or so, then gradually bring her concentrates up to the normal amount over the next ten days. Allow her normal hay and succulent foods throughout. In the older animal too much concentrate ration immediately before and after kidding may well result in milk fever (see page 103). It is essential that only about half the goat's potential production is withdrawn for the first fortnight.

The goat's appetite will increase enormously after she has kidded and, provided the grain ration is increased gradually, her bulky foods can be fed normally. Remember that she will now drink far more water so never fail to check this.

For the first few days, if the kids are being left with their mother, you will only need to ease the overfull udder. A large teat still full of the 'nature' or fullness that develops at kidding may be too fat for the kid to grasp in its mouth. This disappears during the week following parturition. A squeeze or two will relieve the situation and the kid will soon suck strongly. Check the size of these large teats each morning. A small kid may have difficulty with the overfull teat that may have filled over night. After a day or two the kids should be able to manage unaided. This problem may not arise in the goat with smaller teats, but it is always worth checking that the kids really are sucking strongly and frequently.

For several weeks after kidding, the goat will intermittently pass blood and mucus from the vagina. She may need sponging down on occasions to discourage flies. This discharge is quite natural, and only if it appears in the form of excessive fresh blood, or it has an obviously unpleasant smell, need the vet be called to investigate possible retained placenta, an infection or haemorrhage. I always lift the tail and check the goat daily, until the discharge ceases. Such draining is particularly noticeable after the goat has been lying down for some time, and is also particularly noticeable from the fifth or sixth day up to a fortnight after kidding.

The udder, or 'bag' as it is sometimes known, is comprised of special milk-secreting cells and a branching system of ducts. Except when milk is actually drawn, there is a circular muscle which keeps the teat orifice closed. The entire gland and duct system is enclosed in the udder which varies in the amount of fat and other tissues, according to the individuality of the goat.

The udder is easily damaged by mishandling which may result in inflammation, or mastitis as it is known (see page 104). The process of milking by hand should ideally imitate, as nearly as possible, the kid suckling. Watch the kid when it suckles. It moves from the head of the goat down her side, nuzzling and touching her flank, its nose touches the udder and gently nudges it several times before starting to suck. It then gently takes the teat in its mouth and firmly and rhythmically sucks, interspersed at intervals by nudges. As the kid grows older and stronger, the sucking becomes more vigorous and this is often countered by a kick from the goat.

It is best to stick to a regular routine when milking the goat. Tie the animal up firmly, give her the concentrate ration in a bucket, wipe the flank with a damp cloth to lay the dust and also to imitate the young kid moving in. A disposable towel wrung out in warm water is wiped over the udder and teats to clean them. This also approximates to the nuzzling and butting of the kid prior to suckling. An udder cream wiped on the teats and the milker's hands will act as a lubricant and antiseptic, this will also help to prevent chapped teats and hands in winter. The cream can be bought in 1 pound tins or plastic containers. However, it is cheaper to buy a 10 pound bucket which can be used to fill a screw-topped honey jar and kept in the milking shed.

The technique is to milk out the goat as quickly,

quietly, gently and thoroughly as possible. Gently grasp a teat in each hand as shown in Fig. 25, squeeze the teat with thumb and forefinger, the top is then closed and the milk is isolated from the udder in the teat. The second, third and fourth fingers are then closed in rapid succession. This forces the milk into the bucket. The stimulus to release the milk is less effective after the first few minutes so milk as quickly as possible without being rough or appearing to rush. As your technique improves, the milk will froth as it goes into the bucket. A good milker will appear to be almost completely still apart from the hands moving. As the udder begins to empty, the flow of milk will begin to slow up and a little *gentle* massage may help to ease the last and creamiest of the milk from the udder. Any roughness will damage the udder and possibly result in a hoof in the bucket. Avoid grasping the udder as well as the teat. A goat with short teats will mean that the milker will have to employ a stroking action using thumb and forefinger.

The size of the teat is a point to consider when buying the goat; Fig. 25 shows examples. In a goatling it may be difficult for the novice to decide

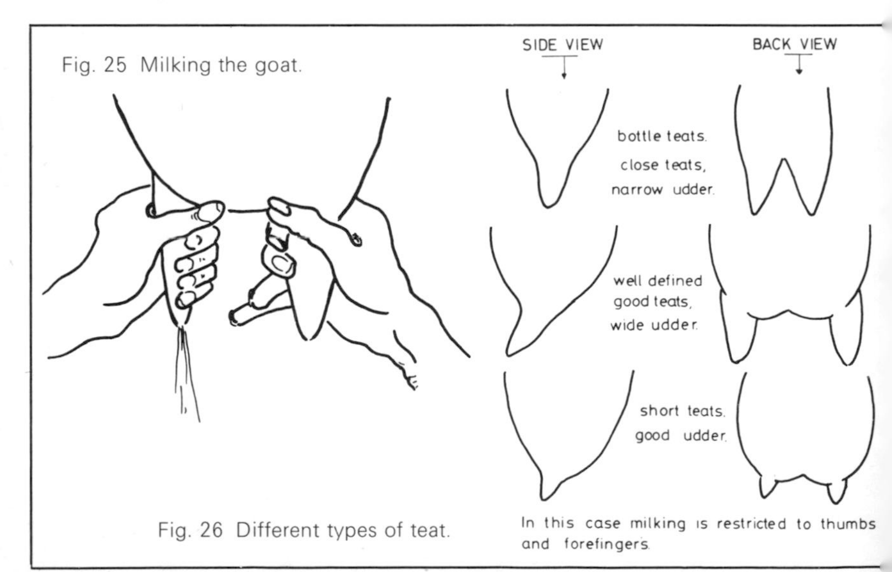

Fig. 25 Milking the goat.

Fig. 26 Different types of teat.

how she will develop. Inspection of the dam or sisters, if at all possible, may give an indication of possible development.

It is a useful practice to direct the first few squeezes of milk into a dark coloured cup. This can give an early indication of trouble. Any milk colour other than milky white can be caused by mastitis. Another form of trouble may show as pink milk; this can be caused by a broken blood vessel in the udder. Clots or strings of milky material may indicate advanced mastitis (see page 104). Learn to judge the warmth and feel of the healthy, disease-free udder. A slight suggestion of extra heat which is barely perceptible may indicate infection of some kind and steps can be taken to remedy this early on. The fore-milk is best retained and offered to a cat or dog or flushed down a drain away from the milking area to reduce the spread of possible infection.

While milking, talk to the goat; she may even appreciate whistling or singing. She will soon put a leg back when you ask her. She will like having a fuss made of her and will enjoy this handling; she will respond by dropping her milk quickly and milking freely.

Points to be observed:

1. Regular milking times.
2. Clean the goat's flank and udder.
3. Use an udder cream to lubricate teats and hands.
4. Check fore-milk in a dark coloured cup, prior to milking, as this will more easily show up any clots due to mastitis.
5. Quiet, gentle, thorough milking.
6. Empty or 'strip' the udder gently.
7. Make a fuss of the goat after milking.

Some people milk the goat from behind (taking care to avoid a bucket of excreta!) but this will depend on the shape of the udder and the teats. If she is milked from the side always milk from the same side so that she knows what to expect. Many goat keepers like to train the goat to stand on a platform for milking (see Fig. 27), the milker then sits on the platform at the side of the animal while milking.

A gallon milking bucket with a wide base is ideal for milking a goat. Plastic buckets are easy to keep clean but do not last more than a year or two; tinned metal buckets are substantial and easily sterilized. However, stainless steel buckets will give many years of service but they are expensive. A lid is advisable to keep the dust from the milk before it is cooled.

Fig. 27 Milking stand.

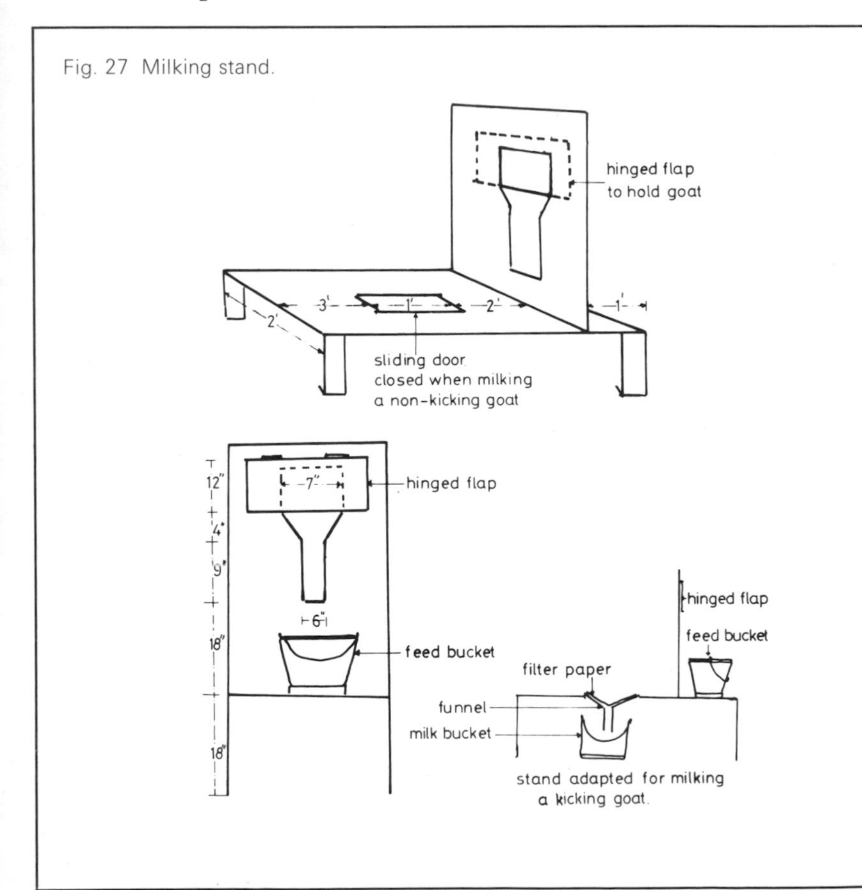

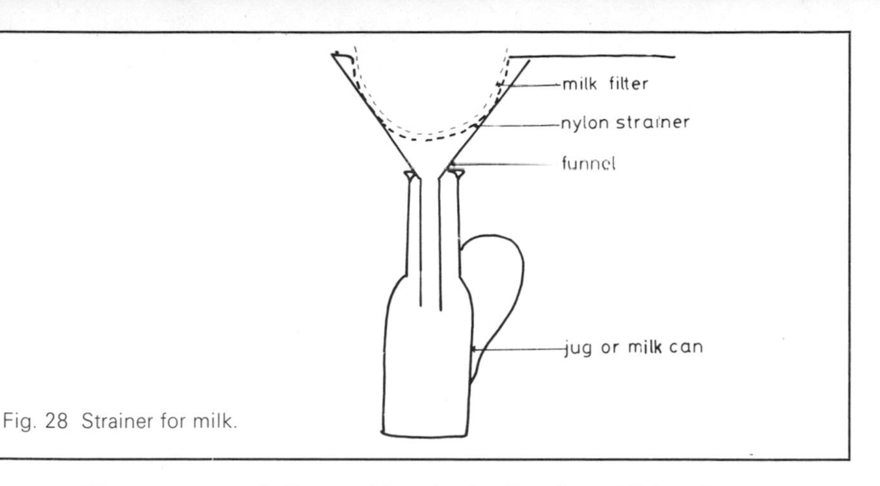

Fig. 28 Strainer for milk.

However carefully and hygienically the milking is completed, subsequent straining and cooling is essential, and should be completed as soon as possible after milking.

Ideally the milk is kept in a refrigerator. If the containers are arranged in a definite order the oldest milk will always be used first. Goat's milk, being higher in lactic acid than cow's, quickly picks up taints and goes sour rapidly if milking is not carried out under the cleanest conditions possible. Tainted milk will quickly put off a newcomer to goat's milk. Care then must be taken to see that once the milk is cooled and in the refrigerator, it is not stood close to any strong-smelling, scented foods. Excessive handling of the milk after milking gives bacteria the opportunity to multiply rapidly. To avoid this, pour the milk from the milking bucket through a filter which stands in a funnel leading directly into a milk receptacle standing in running cold water (see Fig. 28). When cool it can be lidded and stored in the refrigerator.

It is possible to buy milking machines for goats, but they are very expensive. The machine has to be dismantled, washed thoroughly and sterilized after each use, and for the family with only one goat the expense and also the labour involved will not be warranted.

5 Kid Rearing

One of the delights of keeping a goat is the annual or biennial additions to the family. Goats commonly produce twins, numbers may range from one to five, or even six, at birth.

For the squeamish, the culinary delights of kid meat are not experienced. The billies will have been either destroyed by the vet or by the method outlined in the previous chapter.

Billies are sometimes reared as pets, having been castrated while they are kids and before they reach sexual maturity possibly as early as three months of age, but they may not be really suitable. They tend to be bossy unless handled extremely firmly while young, the kid charm is lost all too soon, and you are left with a none too affectionate and rather large pet, which may possibly need far more food than the nanny goat.

The effect of castration is to produce a larger animal than normal at maturity. Where a nanny kid would weigh 100 pounds at approximately a year old and her entire brother 125 pounds, her castrated brother may even weigh up to 150 pounds. This ability can best be utilized to produce *meat* rather than *pet*. Into the bargain, the meat is said to taste less 'goaty' than that of an entire billy, even at the tender age of three months.

Whatever the ultimate purpose of the kids, they will need to be reared. The natural method is to leave them to suckle from their mothers at will. The kids will grow well getting milk of the correct quantity, quality and temperature. They will also learn how to respond to various situations from their dam, and may be initiated into what *is* and what *is not* good to eat by her example. She will, however, after the first fortnight need to be milked out daily or even twice if she appears to have a lot of milk. She will produce too much milk for the kids initially (unless she has to feed three or four kids). The milking will make sure that she is emptied thoroughly daily, which in turn will maintain a plentiful supply of milk. It will also reduce the chance of uneven udder development which selective sucking by the kids can so readily produce. However, kids reared this way *may* not be so bold and confident with humans as those reared on the bottle. The milking goat will also not be producing so much milk for the house.

The kids will have to be completely weaned by six months of age and may resent this strongly, moving heaven and earth to get back to the milk bar and the comfort of their dam. Weaning, at whatever age, is best done over a period of time if the kids are not to suffer a set-back in growth. Whenever it is, the gradual process is vital or the shock to the kid will cause it to lose condition. With autumn approaching the kid can easily fall prey to a heavy worm infestation. Pneumonia may strike the leaner animal already suffering from worms. On the other hand, the over-fat kid, which has been liberally fed on a high concentrate diet, may die from entero-toxaemia after some sudden weather or diet change.

Weaning is best done, initially, by shutting the kids away from the dam at night at about three months of age and, in order to avoid distress to all concerned, the kids and their dam will be happier being able to see each other. A partition as previously described, which is solid up to a height of three feet, with a further two to three feet of horizontal or vertical bars spaced to enable the animals to see each other (but not get their heads caught), will be suitable. Ensure that the kids have an adequate supply of really palatable browsings and hay to pick at while they are in their run, to encourage their digestive systems to develop a capacity for bulk. After a month, gradually cut down the time the kids remain with their mother during the day till, at about five months they are only having one feed a day (possibly at night) from their mother. They will already have learnt to pick at their mother's concentrate ration, but in order to build strong, bulk-consuming animals their grain ration should be confined to about $\frac{1}{2}$ pound twice a day. Continue to maintain their intake of high quality roughage and bulky foods including a little sugar beet pulp, kale and other brassicas in moderation, not forgetting as much hay as they can consume. If the kid's diet has been sufficiently varied, including plenty of the bulky foods, the final break from the mother, at around six months, will hardly be noticed.

Hand rearing, on the other hand, will provide us with exceptionally affectionate and friendly kids. This is an extremely satisfying way to rear kids but it is costly in time and labour; it also requires a fair degree of stockmanship, commonsense and also the ability to spot trouble before it raises its ugly head. Most people allow the kid to remain with its mother for the first four days before weaning it on to a bottle. The goat will need to be milked under the strictest of hygienic conditions. Lid the milk bucket, strain, cool and refrigerate the milk till it is needed. The kid may take a little time to adjust to the new milk dispenser. A baby's bottle, designed to allow air to enter as the liquid is drawn off so that a vacuum is not produced, is ideal; the kid will soon grasp the idea and learn to suck vigorously. After a few days the baby's bottle will have to be replaced by a glass lemonade bottle with a strong calf-teat. The teat will have to be held on by hand however firm it appears to be, as kids can easily choke on a dislodged teat. Figure 29 shows the correct method of bottle feeding.

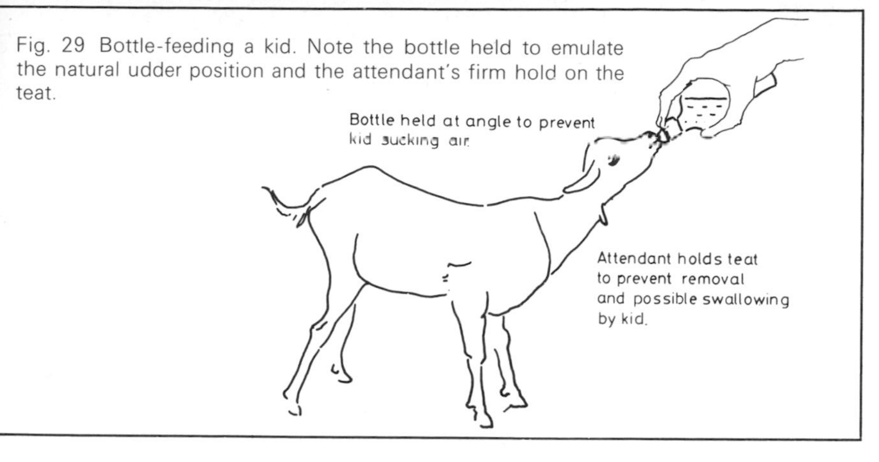

Fig. 29 Bottle-feeding a kid. Note the bottle held to emulate the natural udder position and the attendant's firm hold on the teat.

Initially, the kid will need four to six feeds a day. The bottles used must be scrupulously cleaned between each and every feed. Scours can quickly develop if the milk is at all dirty, or offered to the kid at any other temperature than blood heat. There is a good chance of under or over-feeding, but the former is the lesser evil. The correct quantity will be arrived at by studying the condition of the kid and using the quantities set out in Table 3 as a guide. Over-feeding overtaxes the digestion of the kid. It will lead to an unusually high number of bacteria developing in the gut and the resulting inflammation causes scouring and general unthriftiness. Prolonged malnutrition will result in an animal with the condition known as 'pot-belly'. The animal will have an uncomfortably large belly which will easily be distinguished from the kid with the well developed rumen. The pot-bellied animal will have the appearance of a hairy football on spindly legs. It is easy to succumb to overfeeding the young kid as it consumes milk with great gusto. When the kid has finished it will butt with its head for more, but the temptation to succumb to its wishes *must* be resisted.

Keep the young kids within sight and sound of their mother while she is stalled, and they will soon associate the human with their feed. The mother will happily let her milk down to the human milker while she can still see her kids.

The kids will need a little, very good hay offered to them several times a day to tempt them. I like to leave it on the clean area of the floor of the stall. It may be wasteful but it eliminates the risk of a kid possibly hanging itself in a hay net or hay rack however foolproof it may appear to be.

A bucket with a mere handful of good, plump, crushed oats (not the sample which appears to consist mainly of husk) or coarse, dairy goat ration can be offered after each feed. They will only fiddle with it at first, but by a month old they will be looking forward to it. Do not leave any uneaten grain ration about; remove it as soon as the kids have finished. Food left lying around encourages vermin and sparrows. Goats will rarely go back to food that has been sniffed over and rejected for any reason. Offer a wide variety of bulky succulent foods from an early stage, obviously in small quantities. If there is a rough weedy patch in the garden, the kids may be let out to graze after their milk *and hay*. Avoid turning young kids onto a rich, dairy cow ley containing expensive grass and clover mixture.

Table 3 Bottle feeding for kids

Age of kid	Number of feeds per day	Approximate quantity of each feed depending on size of kid
0–7 days	5	$\frac{1}{4}$ pint
1–2 weeks	4	$\frac{1}{4}$–$\frac{1}{2}$ pint
2–3 weeks	4	$\frac{1}{2}$–$\frac{2}{3}$ pint
3–6 weeks	4	$\frac{2}{3}$–1 pint
6–12 weeks	3	1–1$\frac{1}{2}$ pints

A kid nursery, with clean water available at all times and a shed with its back to the wind, is ideal (see Fig. 30). The size can be dictated by the length of netting available, paying particular attention to straining it as

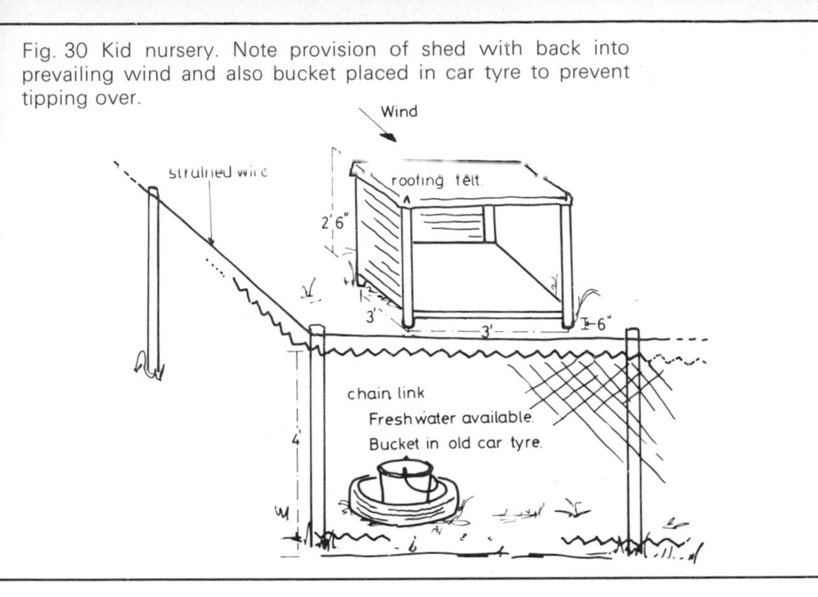

Fig. 30 Kid nursery. Note provision of shed with back into prevailing wind and also bucket placed in car tyre to prevent tipping over.

tightly as possible. A wire strained 9 inches above the netting will discourage the mountaineer.

Do not be worried if, from about one week old, the kids are seen eating soil. Ideally, they will appreciate a freshly turned-up molehill provided it is situated away from any goat-sick pasture. If this is impossible dig some soil known to have been free from goats and offer it to the kids in a box in their stall. A mineral lick should also always be available.

The aim in goat rearing is to develop a roomy animal with a strong skeleton and plenty of capacity for bulk. The fat, soft animal fed on a high concentrate ration will not be so disease resistant and will have little capacity for eventual high milk production; it will also need warmer housing.

If the goat's milk is needed for the house, it is quite possible to gradually change the milk ration over to a proprietory milk substitute when the kids are at least one month old. Gradually increase the proportion of substitute to goat's milk over a week to cut down the risk of a digestive upset. Instructions as to how to make up the powder will be included in the sack.

Follow these carefully. On the other hand, skimmed goat's milk can be gradually introduced at three months, the fat being made up by making sure that the animal gets its ration of crushed oats. Skim-milk feeding requires extra care as kids very easily scour on this product. The feeder must be especially observant to watch the kid's excreta, as any change in its form from the normal pellets will indicate possible trouble. Remember that any digestive upsets may take up to three days to make themselves apparent after the initial irritant has been introduced, although in kids an upset will show itself sooner rather than later.

If an attack of scours should develop, cut out milk feeds for twenty-four hours substituting cool, boiled water with two teaspoons of glucose per pint added. Gradually introduce milk again. Offer kaolin powder on a wet finger to the kid before its bottle. If the animal is so loose that the solid excreta comes out in the form of liquid, obtain professional advice as quickly as possible (see page 108).

After the female kid has survived her first winter and second spring, she will, on reaching twelve months of age, be known as a goatling. The cares and worries of the kid rearer are almost over, but this does not mean that one can neglect her. A wide variety of hay, access to mineral lick, clear clean water at all times, in conjunction with plenty of sun and exercise will result in a fine, well-grown goatling ready to join the breeding herd in the autumn. Her capacity for a large quantity of bulk foods, built up during the goatling stage, will prepare her for a long and productive life as a milker.

At one time breeders endeavoured to breed with only hornless goats in an effort to produce a purely hornless strain of goat which would eliminate the need to disbud the kids at about five days old. It was discovered that mating between two hornless animals sometimes produced a so-called hermaphrodite. This is a goat possessing the reproductive parts of both sexes, neither of which are fully functional. It is more commonly found in apparently female kids. When the kids are born, the female should be carefully examined. If it has an enlarged clitoris at the base of the vulva the size of a pea enclosed in the lips of the vulva, then the animal should be destroyed as it will be sterile or only fit for meat. On the other hand the produce of two hornless parents may *appear* quite normal, and it is not till it has been mated and found to be infertile that it is discovered to be a hermaphrodite. In the male, however, this condition is more difficult to detect. Not until he has been tried with a couple of females or so at six months without success can one suspect that it might be a hermaphrodite. However, mating horned animals will dispense with the possibility of producing this.

A naturally hornless animal has two lumpy protuberances on its head. The head of an animal which has been de-horned as a kid will appear almost flat. At birth, the potentially horned kids will have two little curls of hair around the spot where the horns will develop. The billy kid's horns may already be showing. A veterinary surgeon will disbud kids after a local anaesthetic has been administered. The kid may shake and shiver a little after the disbudding but will soon recover and by the next day, apart from shaking its head intermittently, will have quite recovered. The longer this is left after the horns have broken through the skin, the more difficult it will be, as they develop rapidly.

Goat horns, if allowed to grow, can be something of a hazard and it is therefore safer to have the horns removed. Also, should you wish to sell a kid, a buyer may not be so willing to give you a good price for a kid with horns.

The goat's hooves will need attention and possible trimming from an early age if they are not to become deformed. A kid accustomed to having its hooves handled from birth will accept trimming with little fuss (see page 116). This may have to be done as early as six weeks if the animals have been reared indoors on soft litter.

The various stages of a goat's teeth development are shown in Fig. 31.

Both male and female young are known as kids until one year old. Thereafter the female is known as a goatling and the male a buckling until two years. The adult male over two years is generally termed a billy and the female a nanny, but some goat keepers call them 'adult male' and 'adult female'.

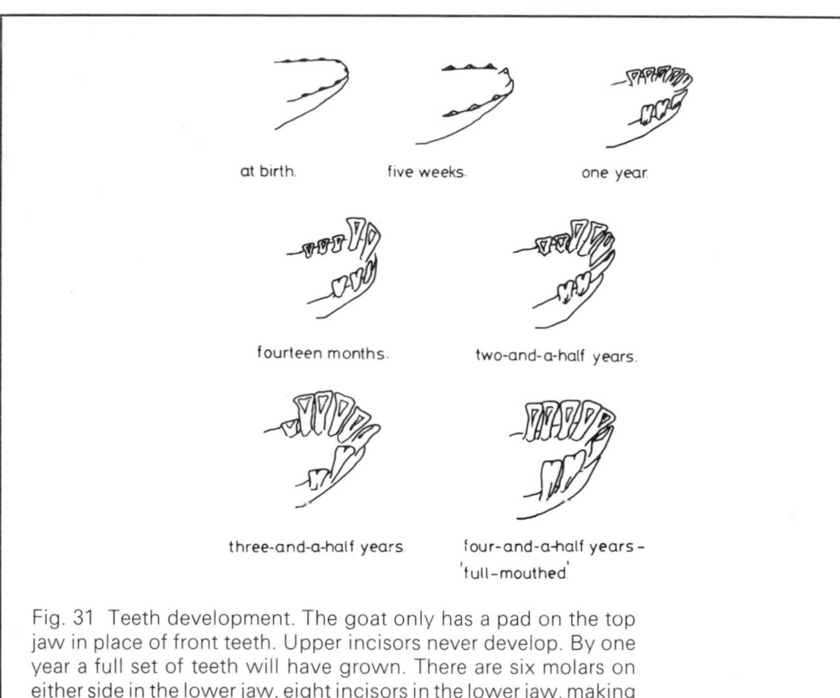

Fig. 31 Teeth development. The goat only has a pad on the top jaw in place of front teeth. Upper incisors never develop. By one year a full set of teeth will have grown. There are six molars on either side in the lower jaw, eight incisors in the lower jaw, making a total of thirty-two teeth in all. At one year the milk incisors are gradually replaced by permanent teeth until the time the animal is about four-and-a-half years old when it is known as 'full-mouthed'.

6 Goat Products

The product which first springs to mind when considering our goat is generally the milk. Milk is obtained from the goat twice a day as described earlier in the chapter on breeding. It is not unusual for a goat to produce 15 pounds weight of milk in twenty-four hours. Exceptional animals have been known to yield over 20 pounds. However, the average goat with reasonable management will produce about 8 pounds a day at the peak of lactation. This is usually reached about six to eight weeks after kidding.

Not uncommonly, goats have the ability to maintain their lactation over two years or more. However, if the animal is left too long before being mated (more than two years) difficulties may be experienced getting her in kid once more.

Some goats, having reached a peak in their production, rapidly fall off. Others may maintain a reasonable level throughout the majority of their lactation. It is not unusual for unmated goatlings to produce milk. I have known goats which have yielded milk for years without ever having been mated, but heavy daily yields are not nearly so likely from these goats.

Milk is an ideal medium for bacterial growth and it cannot be emphasised too strongly that strict hygiene is essential. Goat's dung being normally of a dry nature, will mean that there is far more likelihood of dry dung particles sticking to the goat's hairs in minute quantities. A damp cloth wiped over her sides will help to lay the dust. A concrete-floored milking area, which is hosed down twice daily, is preferable to milking in the living stall. However, milking a goat out of doors is perfectly possible especially if the goat is trained to jump on a milk stand for milking. This stand can be regularly scrubbed. It can also be transported to a suitable place as required. Most gardens have somewhere that is dry and sheltered from the elements (as well as being convenient to milker and goat). There is the added bonus of the purifying action of the sun on the stand which will help keep down the numbers of bacteria.

I have milked out of doors for the last nine years or so, and in spite of the fact that the rain invariably starts at milking time, a well-chosen milking spot can save the expense of building a special shed.

Try to wear a clean overall at milking time complete with hat or scarf; it will keep your own clothing clean as well as reducing the risk of passing on infection.

The milk should be strained, as shown previously, and cooled (a fridge will do) away from strong-smelling foods. In the newly kidded goat, a slightly bitter tasting milk may be produced caused by possible digestive upset following kidding (too many concentrates and not enough roughage). Cobalt salt offered daily will help to cut down the likelihood of taints.

Milking utensils must be rinsed thoroughly in cold water after use and preferably scrubbed. A hot detergent and water scrub will follow. The utensils are then rinsed in cold water and left upturned on a draining board in a cool clean place till required. Unless the scrubbing is done really thoroughly, there is bound to be a build-up of harmful bacteria. Most dairy sundriesmen stock Ministry-approved sterilants and these can be bought in quantities suitable for the one goat family. A gallon of sodium hypochlorite may last a year, larger quantities will lose their potency after longer storage. Follow the maker's instructions carefully. Some advise rinsing the utensils with clear water before use, and others suggest milking directly into the bucket freshly emptied of sterilant.

Disposable kitchen towels for wiping the goat are ideal as they save the inevitable daily boiling of cotton udder cloths. Disposable milk filters are essential if you are going to produce clean milk consistently; these can be obtained from a dairy equipment supplier.

Cream

Goat's cream is pure white, less 'greasy' than some cow's cream and is also very much easier to digest. It whips to a greater bulk than cow's cream and is therefore more economical.

Fresh Cream. Unless there is milk available from at least two goats (when a cream separator may be used) it is best to obtain cream by pouring the warm, freshly strained milk directly into a newly scalded shallow bowl where the milk comes within an inch or two of the rim. After thirty-six hours in a cool place, the cream will have risen to the top and this can be skimmed off with a skimmer (a shallow spoon with holes as shown in Fig. 32). The milk will pass through the holes leaving most of the cream on the spoon. It is wisest to stand the bowl in a refrigerator or a very cool larder as this helps the cream to harden somewhat. As the remaining skimmed milk has a certain amount of fat remaining it can be used as whole milk or fed to older kids.

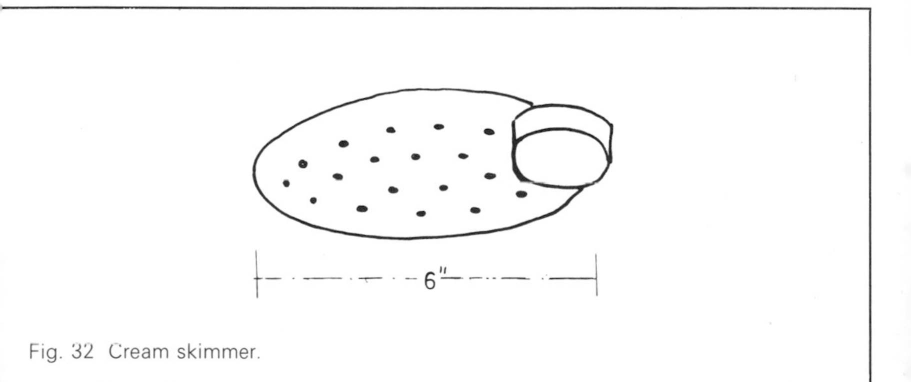

Fig. 32 Cream skimmer.

New cream separators are now available once more. However, you may be able to obtain an old hand-operated (see Fig. 33) or electric model much cheaper at a farm sale. Separators are a very efficient method of cream making. Our first separator was given to us covered in owl's droppings and straw in 1965. It had not been used since the end of the war and was, we discovered, so old that spare parts had ceased being made in 1936. We removed all the dirt and grease, exposing its maroon paint with gold beading. The time came to experiment with a mixture of oil, water and detergent. While my husband turned the handle, I poured the mixture into the receiving pan. After a few turns we were the proud producers of cooking oil in one receptacle and water and detergent in the other. The rubber washer it required had to be adapted from a length of rubber beading obtained from a local engineering firm as the correct washer was unobtainable. The cream screw will need to be tighter than that setting used for cow's milk, as goat's milk has very small fat globules. When the separator is as venerable as ours, cream thickness can only be controlled by the speed at which the handle is turned. For satisfactory separating, the milk should be as near bloodheat as possible. The minimum amount of milk required in order to avoid wastage from cream remaining in the machine, must be about a gallon (the morning milk from two goats or, heat the milk from the previous milking to bloodheat and separate with the fresh milk). Do not separate milk from the freshly kidded goat but wait for at least five days when the colostrum will have ceased. (The colostrum, on the other hand, can be used to make superb baked custard. The freshly milked colostrum is put in a greased baking dish and cooked in an oven at 300°F for about forty minutes or until set. The finished dish tastes like a rich egg custard.) The skimmed milk from a separator

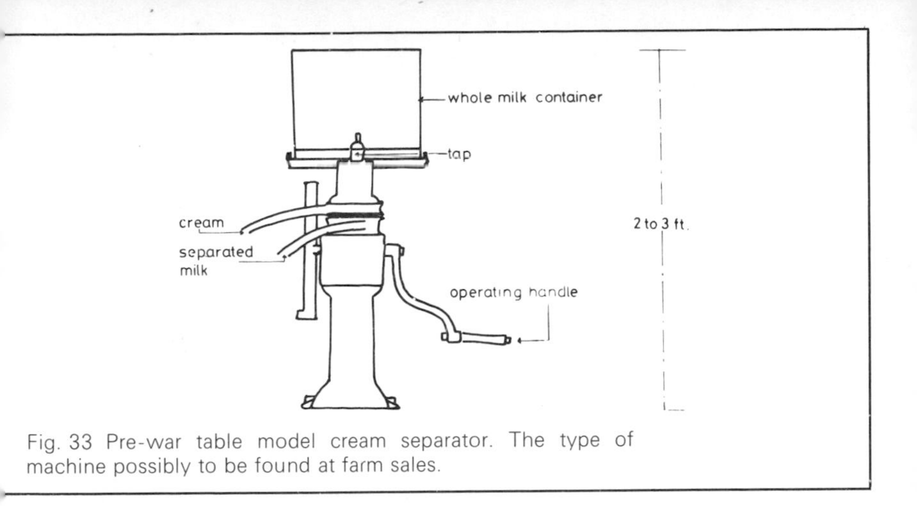

Fig. 33 Pre-war table model cream separator. The type of machine possibly to be found at farm sales.

can be used normally, but it burns more easily when boiled than does whole milk. You will also find that as the grasses come into seed in summer, and in autumn as dry feed increases, the cream thickens considerably. The screw on the separator will then have to be adjusted accordingly or the resultant thick cream will completely block the outlet.

The machine must be rinsed thoroughly in cold water immediately after use, washed in warm detergent and water, and dried. We put the rinsed, washed parts in the *coolest* oven (*c.* 175 °F) of the Aga to dry. If put in any warmer oven the soldered joints will melt with disastrous results, as we learnt by bitter experience.

Clotted Cream. The making of clotted cream is a longer, more complicated process, but it is well worth the trouble if the characteristic flavour of the product is preferred; it also has the added advantage that it keeps longer. The new warm milk is strained into a shallow pan and left in a cool place for up to thirty-six hours. The pan is then set on an asbestos mat over a cool heat (the simmering plate of a solid fuel stove is ideal) till the fat begins to look oily and forms a crust (after about one and a half to two hours). The temperature may then be about 170–190 °F. When the crusted cream is drawn back with the tip of a spoon, a little cream should be apparent underneath. If there is no flat pan available the milk can be set in a shallow bowl and stood in a baking tin of warm water on the simmering plate. The pan or dish is then cooled as quickly as possible and the resultant cream removed about twelve hours later. This method of cream removal is rather more efficient than the hand method given for fresh cream.

Yoghurt

Our surplus skimmed milk is used for making yoghurt. The milk is poured into a Pyrosil dish (or you can use a saucepan) and left on the simmering plate of the stove (an asbestos pad on an electric or gas stove turned down low will do) for about an hour and a half. The resulting temperature inhibits the growth of bacteria and reduces the quantity of milk a little. Having bought a fresh carton of plain, unsweetened yoghurt (we like to use the one made with goat's milk, though cow's yoghurt will do), the container of milk is cooled in a bowl of cold water to just over 105 °F (one learns to judge this by the feel). Two tablespoons of yoghurt are stirred into approximately two pints of warm milk. We then leave the milk and yoghurt mixture in a covered Pyrex or Pyrosil pan (a metal saucepan is inclined to flavour the yoghurt) on a folded newspaper on the back of the Aga. In cold weather it needs to be wrapped in a clean towel to help maintain the correct temperature. An airing cupboard can be uscd, but a little experimentation will be needed to find the ideal position and temperature in each house. If no suitably warm place is available, the milk and yoghurt can be put in a pre-warmed, wide-mouthed, vacuum flask. The mixture is then left till it thickens, which may be any time between three and six or more hours. Cool it rapidly before use. During the winter, yoghurt left to make overnight works well, but in summer it may curdle if left for twelve hours due to the warmer surroundings. A special yoghurt culture can be obtained from some dairy suppliers, but it is very often more convenient and cheaper to use the yoghurt on sale at the grocers. There are yoghurt making machines available which consist of an electrically heated, thermostatically controlled, insulated container which holds a plastic lidded vessel. The preheated milk and yoghurt mix (at approximately 105 °F) are put in the plastic vessel and the machine plugged into the electricity supply. The yoghurt will have to be watched as the container does not automatically switch itself off when the yoghurt is made. It is useful where there is no suitably warm and clean place to make yoghurt.

If the milk used is too hot, a clotted, rubbery, stringy brew will result which is only suitable for the pig. Milk not warm enough, on the other hand, will take a long time to turn into yoghurt and will possibly taste bitter due to the action of other bacteria.

Retain a small quantity of yoghurt to make the next lot. This reinoculation can continue till the yoghurt becomes acid and bubbly or makes a lot of whey when stirred. Some like to make a thicker yoghurt by adding two tablespoons of dried milk to each pint of goat's milk.

Experimentation is needed, don't be put off by failure at the beginning. Before long you will be making superb yoghurt tasting far better than any bought variety. You never need be at a loss as to what to have for a pudding. The addition of fresh or tinned fruit, nuts and raisins will be very popular. For the glutton, cream and added sugar will be a further delight.

Yoghurt will not be able to be made if the goat is receiving any penicillin or other antibiotic treatment to the udder. A couple of days or so will have to elapse before the milk is suitable for yoghurt once more.

Cheese

You may well find that the yoghurt production catches up with you, the refrigerator and all available cupboards will be filled with yoghurt of varying ages. This is the time to start experimenting with a simple cheese-making process. The British Goat Society publish an excellent booklet called *Dairy Work for Goat Keepers* priced very reasonably indeed. It is full of detailed information on the making of all types of cheese, besides other valuable advice. To many newcomers to goat keeping, the processes involved will appear time-consuming and complicated, so a simple soft cheese which needs little equipment for its production can be the introduction to the more involved types.

Cheeses can be divided into two basic types:

1. The true hard and semi-hard cheese. This is made from milk with the addition of rennet to form the curd. Cheese made by this method will require a ripening period before being ready for consumption.
2. The soft cheeses which develop their own acidity without the addition of rennet. These cheeses must be eaten fairly quickly after they are made as they have a limited storage life in a refrigerator of about two weeks (this depends on the cleanliness of the milk production!).

Yoghurt cheese comes into the latter category. The ready-made yoghurt, having stood for twelve hours in the refrigerator, is poured into a sterilized cheese cloth (an old boiled pillow-case with holes efficiently patched can be utilised). This is hung over a bucket to catch the whey and is left to drain in a cool, clean, insect-free place for about twelve hours. The curd is then scraped from the cloth into a clean mixing bowl, mixed and seasoned to taste (pepper, salt and any other flavouring such as chives, sage, etc. can be used). I then shape it into a rough parcel, re-wrap it in another clean cloth and press it between two plates with a 2 pound weight on top for an hour or two. It is then turned out on a plate for use. It does *not* appreciate being stored in an air-tight plastic container. A china bowl with a saucer on top is more suitable.

The cloths must then be thoroughly rinsed in cold water to remove any remaining curd, boiled for ten minutes, hung in the sun to dry and put away in a clean bag for subsequent use.

The whey can then be made into a cheese which originated in Norway. It is a time-consuming process, but can be made while doing other cooking without too much effort. Pour the whey into a thick saucepan and boil it slowly till it is reduced to about half its volume. Then leave it in a warm place to evaporate for a further twelve hours or so (to fit in with whatever you are doing—it is a very good tempered product). The whey is then boiled again very slowly, stirring all the time till it becomes pasty. It is removed from the heat and beaten well with a wooden spoon till it thickens. Spoon the paste into a greased cup for use. It *can* be shaped and put into a butter paper. If this 'cheese' is boiled too vigorously and over long in the second stage, a toffee-like substance is produced which is deliciously acid, but can only be eaten in saltspoonsful by the very brave. The correct consistency is rather like a firm peanut butter. It can be spread on bread in the same manner as Marmite. This cheese is an acquired taste perhaps, but once you do develop a liking for it, you will never want to be without it. It is of course very rich in protein and will appeal especially to those who cannot bear to waste anything.

A soft cheese can be made with naturally soured milk. The warm milk from the goat is left to go sour keeping the covered, strained milk at about 80 °F. Continue the process as for the manufacture of yoghurt cheese. The chances of unwanted bacteria developing are very much greater in this type.

Lait Battu

This is a very pleasant product which is delicious served with stewed fruit. Fresh, absolutely clean milk is allowed to go sour. Before too firm a curd has developed, strain it in a cloth. The curd is then removed before it loses too much whey. It is beaten with a fork to a smooth creamy consistency, and a little sugar is added. Served with fruit it is very refreshing, being not so rich as cream and slightly acid.

Butter

The cream from one goat will barely be sufficient for butter making as, very approximately, a pint of single cream is obtained from a gallon of milk using a separating machine. Once goat's butter has been tasted, though, you may wish to keep the household short of milk in order to make this delectable product.

Old-fashioned, wooden butter churns are not the necessity which might be imagined, and, in fact all that is really needed is a mixing bowl, a wooden spoon, a dairy thermometer and a pair of butter bats or 'scotch hands'. These can be bought very reasonably from many hardware stores. Butter can also be made with a food mixer working at about ninety revolutions per minute (the slow speed) with a cake-mixer head attached. For those who yearn for a butter 'churn' there are glass models with wood or metal

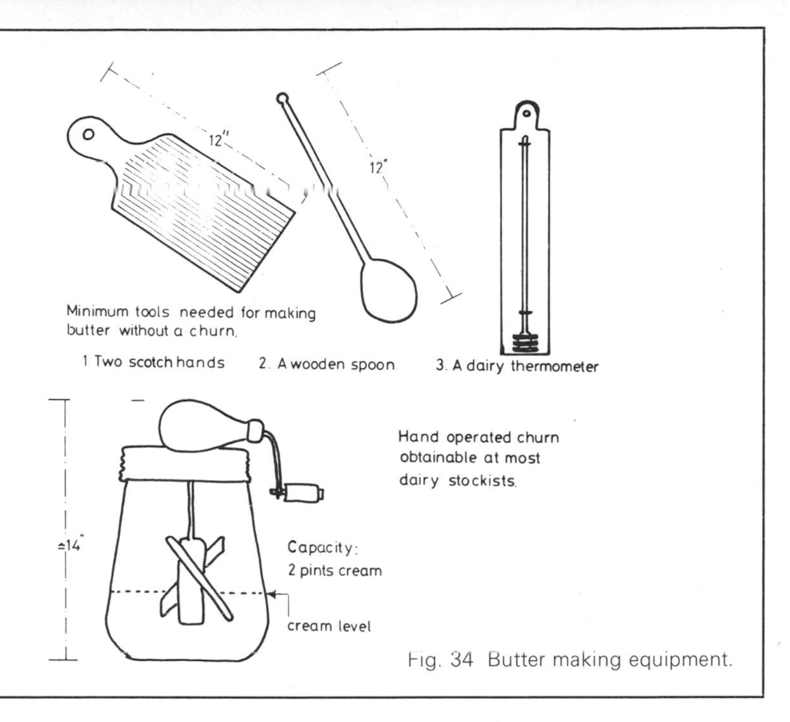

Fig. 34 Butter making equipment.

beaters which can be turned by hand or electricity. The smallest Blowbutta churn will take two pints of cream (the churn should not be filled more than one third full). Figure 34 shows the basic equipment for butter making.

There are three types of cream used for making butter:

1. *Fresh cream.* This can be stored in the refrigerator adding the same quantity of fresh cream daily (and stirring well after each addition) until there is sufficient quantity to make butter. The cream will develop a pleasant, slightly acid smell. A bitter tasting cream caused by dirty conditions is unsuitable for making butter. If this method is used, do not add fresh cream less than twelve hours before making the butter.
2. *Clotted cream.* This cream makes a particularly pleasant tasting butter.

3. *Soured or ripened cream.* This calls for a high degree of hygiene in the milking and subsequent storage. The cream is stored at a temperature of 50–60 °F at which heat the lactic acid sours the cream. If stored in cooler conditions, the cream may develop bitter or bad flavours and the resulting butter is most unpleasant. As above, the quantity of cream added daily needs to be constant and none must be added less than twelve hours before butter making commences.

The only essential tool needed is a dairy thermometer, as the cream must be the correct temperature for churning:

52–60°F in hot summer weather
58–66°F in cooler and winter weather

In order to get the cream to the desired temperature, stand the cream container in a bowl of cold or warm water till the correct temperature is reached. If these temperatures are not observed fairly carefully, the butter will take a long time to make and will not be a good texture.

Paddle the cream regularly back and forth with a well scrubbed wooden spoon (or your clean hand) in a deep bowl (I find this works better than round and round for some reason). Treat the cream gently and firmly. I paddle in time to music with a regular beat if there is any on the radio (brass band marches or Johann Strauss waltzes are good, for example).

After a time the slop of the cream will change its note and the consistency of the cream will alter quite suddenly. At this stage add a quarter of a pint of water for each one and a quarter pints of cream. This water will need to be cold if the weather is warm, but if the temperature of the room is 58 °F or less, the water added will need to be 60 °F. Continue to paddle and gradually the cream will become granular. As these granules develop to the size of small peas, stop paddling and strain off the butter-milk (keeping this for bread or scone making), then add successive amounts of cold water, paddling a little and straining it off till the water is *absolutely* clear. This is essential if the butter is to keep at all well. My husband, on the other hand, loves rancid butter (caused by incomplete washing) as it reminds him of the butter made by the dairy girls when he was a student! When the granules are thoroughly washed, sprinkle about a dessertspoon of *dairy* salt for each pound of granules and at this stage add Annatto if a golden yellow butter is desired. Annatto can be obtained from dairy suppliers; on the other hand, one can use small quantities of saffron, or the juice of grated raw carrots or marigold petals.

The butter is now ready to be worked on a thoroughly scrubbed board kept solely for butter making. Scotch hands (wooden pats preferably made of box wood which does not splinter easily) are used here to work the butter in an effort to expel all the water. Slope the board slightly to assist the water to drain away. This can be done on a draining board perhaps, so that the drainings will run into the sink. Obviously this process will be more easily completed in a cool place in hot weather. Leave the lump of butter for twelve hours before making it up into quarter or half pound pats. It is fun experimenting with patterns on the top of the butter made from different pats with the scotch hands. The wood whittler will perhaps enjoy making a particular boss to mark each portion. I like to make a few butter balls or curls each time the butter is made. As they are fashioned with the wet pats, put them into slightly salted water till enough have been made. Shake them dry and then freeze, arranging them slightly apart in a polythene bag and lay the bag flat in the freezer. When frozen they can be thrown together as they will no longer stick. These are very convenient for an emergency as they thaw quicker than a quarter pound block of butter. The blocks of butter can be wrapped in greaseproof paper or parchment. If your early efforts disappoint, don't be put off, for some curious reason the texture of the butter improves as one becomes more experienced.

There are times, however, when the butter just *won't* 'come'. The difficulty can be caused by the fact that the goat is reaching the end of her lactation, or more commonly, that the cream is not at the correct temperature. In winter it may well have to be brought up to 68 °F or you will still be paddling back and forth after half an hour's hard labour.

Meat

If the goat has bred one billy kid, or worse, two or even three, and you have decided to augment the family meat ration with kid meat, it is essential to brainwash everyone in the household as soon as the kids are born. In fact it is often the adults in the family who are more squeamish than the children. Take it for granted that Billy will make some super joints when he is big enough and the family will accept the fact more readily. It is surprising how a child faces up to the situation when presented with the facts in an unemotional manner. The situation is eased if one is lucky enough to have a nanny kid which will still be there to play with long after Billy has been housed in the deep freeze.

Goat meat can be divided into three types: (1) kid meat, from an animal killed before reaching six months of age; (2) meat from the older castrated male; and (3) meat from the old or barren milkers.

The meat from a three-month-old billy is probably that which will best fit in with the management of the one goat family. After three months he will have to be castrated or he will be a nuisance to any of the female goats on the establishment and may cause unwanted pregnancies. At three months you will also have a carcase which is worth using. Under this age, the carcase may produce little more meat than a large rabbit or a well-grown hare.

To arrange for a kid to be slaughtered contact your local slaughter-house and find out if and when it will be convenient to bring the kid for slaughter. They will usually like to have the animal for a few hours before slaughter so that the gut is reasonably empty. Arrange to pick up the 'pluck' the same day as the animal is slaughtered, this will consist of the heart, lungs and liver. Many slaughter-houses have finished work by lunchtime (as they start extremely early in the morning) so find out what time will be convenient to them.

The best quality skins will be those from winter killed kids which have been reared extensively out of doors. Contact the local hide and skin brokers (the name will be found in the yellow pages of the telephone book). Arrange for the skin to be cured. The slaughter-house may well arrange to have it collected with the other cattle and sheep skins, but do not omit to telephone the brokers, warning them that your skin will arrive on a certain day and that you want it cured for yourself. If this is not made clear to all concerned, the slaughter-house will collect the cash, and the skin will be irretrievable.

If, on the other hand, the skin is to be cured at home, collect it from the slaughter-house as soon as possible after killing. Scrape off all the flesh and fat with a blunt knife, then soak the skin in a bucket of rainwater for twenty-four hours changing the water twice. Dissolve ½ kg of common salt and ½ kg of alum in about 20 litres of warm water in a plastic container. Leave the skin in the liquid in a warm place for up to five days remembering to give it a stir when you think about it. Then remove the skin and spin dry but *do not* rinse. Now rub in a mixture of one part egg yolk to three parts warm water to the skin side leaving it for three days. Hang it to dry in an airy shed and wriggle the skin each time you pass. The leather will appear under your fingers as the skin dries.

Work the skin thoroughly with the hands to keep it pliable, maul it and pull it about till it is quite dry. Stretch the skin on a board, hair side down, mark with a pencil and cut to the required shape with a Stanley knife or special leather knife.

It will be convenient for the carcase to hang at the slaughter-house in their cold room for a couple of days. The slaughter-house may even joint the animal for a small sum. If the carcase is hung for less than two days it will have a gluey texture.

A three-month-old kid may produce, approximately, a 30 pound carcase (about 20 pounds of meat). This can be divided into:

2 legs
2 shoulders
A saddle (rather than chops)
Pieces—best end of neck
head
trimmings
Belly
2 kidneys with surrounding suet

The liver is surprisingly large and exceptionally delicious. The heart is very sweet and tender. If the meat is to be deep frozen, joint the carcase, wrap in polythene bags, seal and remember to label with the date and name of the joint.

I doubt if the older castrated kid is worth keeping for meat unless there is plenty of spare milk and a good, plentiful supply of winter keep.

The old goat will supply little meat, but if hung for a week at the slaughter-house and then cooked as for mutton, with plenty of tender loving care and basting, it will give pleasure in the end. Its skin will need longer in the tanning mixture, possibly seventy-two hours as opposed to the forty-eight hours for the kid-skin.

Kid meat tastes delicious if cooked as for lamb. Remember, though, that there is little fat so it will need plenty of basting if roasted. Allow twenty minutes to the pound in an oven at 340 °F. The breast can be stuffed as for lamb and then rolled and baked. The pieces make an excellent casserole but some people prefer them made into curry. In fact most recipes which use rabbit or chicken meat are suitable as well as the lamb recipes.

7 Accident and Disease

Things can go wrong even in the best regulated circles. It is as well, therefore, to be aware of what to avoid, how to avoid it and the necessary action to take if and when something does happen.

The well-managed goat is less liable to illness and accident. She will be adequately loved, housed, handled, and will have a sufficiency of the foods which will maintain her in the best of health.

One soon learns to recognise the healthy, happy goat. She has a generally attentive attitude with a bright, interested eye, firm mouth and nostrils slightly dewy. Her coat is smooth with a healthy bloom, her skin soft and pliable. She stands four-square with a strong, firm back. She is eager for food and when not eating, happily chews the cud either standing up or lying. She may bleat for attention when you pass, perhaps once or twice if she is hungry and needs moving, but soon settles happily when her needs have been fulfilled. Her tail is held jauntily and the anus is clean and free from excreta.

Feel the ears and neck to learn her normal temperature; often the slightest impression of abnormal heat or coldness when handling her will be the first indication that all is not well. The healthy, happy goat has a pleasant, almost milky smell.

One of the first things to be affected in an animal not in the best of health is her appetite. She may not be cudding (they normally cud for a total of about eight hours in the twenty-four). She may cough and hang her head lower than usual. Abnormal dung will be either extremely hard or loose and unformed. The healthy goat passes firm round pellets which resemble those of a rabbit.

The sick animal often pants rapidly and the coat has an abnormal appearance being harsh, staring and dull. The animal may be restless and bleat intermittently for no apparent reason, standing with her back arched and showing little interest in what is going on around her.

The normal temperature of the goat is between 102 °F and 103 °F. Temperature can be taken with a clinical thermometer smeared with vaseline and gently eased into the rectum while an assistant holds the animal's head and reassures it. The observant goat keeper will be aware of the goat's high temperature without a thermometer, as the goat's eyes will appear glassy and her neck and ears will feel too warm. It is little use feeling the back as only a very high temperature will be felt there. Respiration is normally barely perceptible and is only really obvious in the flank as the animal lies down. The rate can be determined by counting for a timed period and should be approximately eighteen to twenty-five exchanges per minute in the resting animal. Abnormal respiration is manifest by rapid breathing and panting, and this combined with a high temperature indicates the need for a veterinary surgeon as soon as possible. The normal pulse rate, which can be felt below the jaw, should be between seventy and eighty pulses to the minute; this may raise to about 115 pulses per minute or even higher, in an animal with a high temperature, or, of course, after harsh exercise.

The sick goat needs constant loving care and attention. She can easily lose the will to live if she is feeling sufficiently poorly. Where possible a dry, sunny, well-littered box where she can be kept quiet is ideal. It can be convenient if it is sufficiently near the house to be under the goat keeper's eye.

In winter a rug will keep a sick animal warm. If there is electricity available, an infra-red lamp will maintain a comfortably warm sick-bay, but it must be placed high enough so as not to burn the back of the animal when standing. When applying a rug, remember to tie it sufficiently tightly so as not to be a hazard to the goat as it tries to get up, and not so tightly as to cut into her. In an emergency, a hessian sack, split down one side and put on with the remaining corner over the goat's tail, and the mouth over the neck and shoulders, makes a good stop-gap. A sick goat, when recumbent, may possibly need propping up with straw bales. An animal left to lie with legs straight out to one side will be loth to find the energy to stand up, whereas an animal which is propped up on her sternum will need but little assistance to get onto her feet once more. Remember that she will need moving at least four times a day to avoid bed sores and also the likelihood of bloat brought on by the animal remaining in one position which may inhibit the normal expulsion of gases. The goat which is rugged for warmth will need the garment removed at least once every twenty-four hours. Rub her coat gently to improve circulation and shake the rug well before replacing it. Check for any possible pressure points which may develop into sores, e.g. point of elbow, hock, sternum, shoulder, etc.

Tempt the goat to eat by frequently offering her a variety of her favourite foods in very small quantities. Never leave any uneaten food before her after she has taken what she wants. Remember she will need to drink perhaps more than usual. Warm the water and offer it to her in a scrupulously clean, taint-free bucket. Avoid feeding slops as goats don't like this form of food. Some may drink milk from a bowl when all other food has been rejected, an egg or even a spoonful of honey beaten into the milk will sometimes tempt the sick goat's appetite. Remember anything, within reason, is better than nothing passing her lips for long periods. I find a spoonful of honey placed at the back of the tongue, while the throat is stroked to encourage swallowing, appears to help in some cases.

Accidents

Cuts. The healthy goat heals remarkably well when the skin is cut. Long scratches on the udder will knit together within days, provided that the cut is superficial and a milk duct is not involved. Flies must be kept at bay during warm weather. To promote healing apply Acriflex ointment which can be obtained from the chemist. It will need to be applied daily till healing is complete. If necessary, the vet will be able to supply a powder which will keep the wound free from flies and bacteria. Where a milk duct has been severed, the milk will be draining out and the attention of the veterinary surgeon should be sought as quickly as possible. The wound may drip milk for a while after being sewn up. If milk is still seeping out through the wound after forty-eight hours, further action will have to be taken. Keep the animal as quiet and still as possible to aid the healing of the wound. She will be shocked to a certain extent and may not feed and cud normally for a day or two. Plenty of green food will assist healing.

Eyes. A foreign body in the eye may cause inflammation. A squeeze of Golden Eye ointment into the corner of the eye will reduce any discomfort. Hold the eye closed for a moment or two to help the ointment to disperse. The commonest foreign body in the eye is an oat flight which is adherent to the surface of the eye and can be very difficult to remove. This may well require professional assistance. Ulceration of the cornea may result from scratches by brambles etc.

Fractures. These are best dealt with by the vet. Goats are remarkably tough and will seldom suffer a broken limb. If a broken bone is suspected, keep the animal as still as possible till professional advice is at hand.

Poisons. As already discussed, the tethered animal is more likely to succumb to plants which are poisonous. A goat which eats a poisonous plant on an empty stomach is more likely to be adversely affected than one which has eaten a leaf or two or has an already half-filled stomach. The most common poisonous plants are:

1. *Ragwort*—An animal which has eaten this plant will gradually lose condition. It is not normally eaten while the plant is still growing as it has a strong smell and tastes bitter. The more common form of this poisoning is caused through eating the dried plant in hay, the bitter taste does not then seem to be so apparent. A good diet, high in carbohydrate with added sugar or molasses, will

help the animal to recover. If ragwort is not wanted in hay the plants are best hand pulled out of the pasture while they are in flower but before they seed. Take care not to touch your face when the plants have been handled as after an hour or so of ragwort pulling the poisons can make a human feel sick and suffer from a headache. It is best done in small doses over several days!

2. *Wild Clematis*—All parts of this plant are poisonous. The animal will scour severely after eating it. Call the vet immediately.
3. *Laburnum*—The seed pods are especially poisonous. The animal will be in obvious pain and her breathing will be rapid.
4. *Laurel*—This contains cyanide. An animal eating laurel will be in great pain showing signs of increased respiration. Death is usually fairly quick.

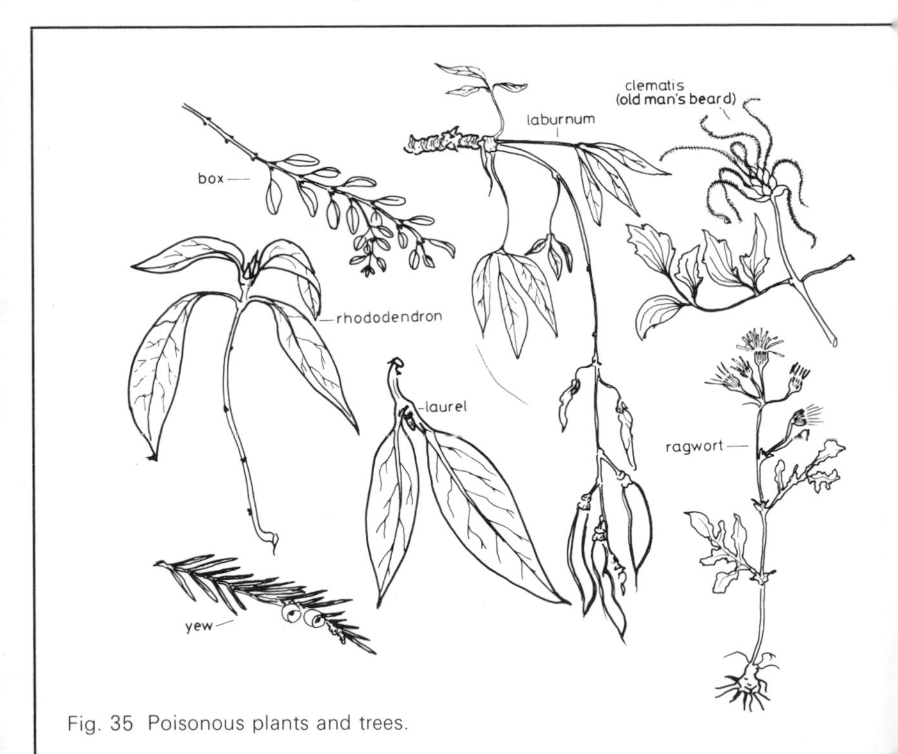

Fig. 35 Poisonous plants and trees.

5. *Rhododendron*—This plant will cause the animal to vomit, signs of rhododendron leaves will be apparent in her saliva which will aid classification of the poison. Stimulate the animal with a drench of alcohol (4 oz of whiskey or brandy, or strong coffee if alcohol is not available) and keep her quiet till the vet comes.
6. *Yew*—Death will result very quickly after an animal has eaten yew. It is very palatable to goats at any time of the year and they will eat it even when they are not hungry. They must not be allowed access to this tree at any time. The berries are especially attractive but only the seed is poisonous.
7. *Box*—After eating this the goat will appear to be in intense pain, she will sweat and salivate freely.
8. *Privet*—This commonly found hedge shrub is particularly attractive to the tethered animal and when eaten, even in small quantities, *can* prove fatal.

All the above plants (except privet) are depicted in Fig. 35.

In all cases of suspected poisoning a drench of one and a half tablespoons of Epsom salts dissolved in a quarter of a pint of warm water must be administered at once, while someone calls the vet immediately informing him of the suspected poisoning. Egg whites may also help. The method for drenching a goat is shown in Fig. 36.

Lead poisoning is commonly found in young kids especially when they are teething. Kids are extremely inquisitive and like to explore with their mouths. An animal only needs a flake or two of lead paint off an old manger, a door or exterior paintwork and the damage is done. The animal will be found with its head turned over its back prior to convulsions. Four to six beaten eggs can be administered as a drench while professional aid is summoned.

Fig. 36 Drenching the goat.

In all cases of poisoning whether plant (except oak leaves) or heavy metal, a good dose of Epsom salts can do no harm. In oak poisoning, where constipation is the presenting sign, liquid paraffin is indicated. Oak leaves in very small quantities may not be dangerous but can help to bind up an animal which is a little loose.

Disorders

Lameness. Regular trimming as described on page 115 will prevent lameness due to overgrown hooves. Foot rot, which is an infection developing within the hard shell of the hoof, may arise in an untrimmed hoof when a smelly discharge will be noticed. Should it be discovered, trim all of the loose horn and expose the infected areas. Paint with a 4 per cent Formalin solution obtainable from the chemist. Repeat weekly till the discharge has ceased. Clean, dry housing and dry pastures will help to prevent this. Lameness may also be caused by rheumatism which develops in older goats kept on draughty, damp floors. Warm, well-littered, draught-free housing will help combat this.

External parasites. Skin parasites are seldom found in the healthy goat kept in clean surroundings. Regular gentle grooming with a dandy-brush, followed by dusting with Gammexane or B.H.C. will control lice. Turn the hairs back and with the other hand dust along the backbone of the goat from behind its ears to its tail. If the goat has a harsh, dull coat, rubs frequently and has developed bare patches, suspect a lice infestation. The most common time for lice to strike is in February when the animals have been indoors a lot of the time. The use of barley straw seems to contribute to lice infestation. Keep a sharp lookout for lice from Christmas onwards, dusting regularly will nip a potential infestation in the bud. Regular creosoting on all wooden parts of the goat's accommodation each summer will help combat external parasites.

Internal parasites. Where goats are continually grazed on a small area the worm burden is bound to build up. Rotational grazing will help to overcome the problem. Tethering the animal on land which has been rested from goats for up to six weeks is ideal as this helps to break the parasites' life cycle. Alternate stocking of the land with other types of animal, e.g. cow or horse will help, but this of course is not really possible in a small area such as a garden. Land which has previously been grazed by sheep is best avoided as goats share many types of worms with this species. It is normal for a healthy goat to live with a light infestation of worms, but the burden rapidly builds up when the animal suffers from bad feeding and generally poor management.

A sample of fresh droppings sent to the veterinary surgeon for a laboratory egg count will determine what types of worm are present and the degree of infestation. He will advise the correct treatment.

Thibenzole is used by many goat keepers for

control of worms. It is the safest method and obtainable in powder or tablet form from local goat clubs or the British Goat Society. To administer the tablet place it on the back of the goat's tongue, gently close the mouth and then stroke the throat. This will help the animal to swallow. On the other hand, the powder can be administered dissolved in a cupful of water and given as a drench. Choose whichever form of administration is found to suit both goat and administrator.

Suggested doses—

$\frac{1}{4}$ tablet for three month kid
$\frac{1}{2}$ tablet for six month kid
1 tablet for one year goatling
1–1$\frac{1}{2}$ tablets for adult depending on weight

An animal with a rough coat, poor appetite and inclined to have loose droppings must be suspected to be suffering from some type of worm infestation.

Dosing a goat heavy in kid with Thibenzole is permitted as long as she is handled gently. A worm dose recommended by the veterinary surgeon and administered a week or ten days before the goat goes to the billy is a good plan.

Coccidiosis usually occurs where many goats are kept together in damp conditions. Symptoms of coccidiosis in goats will be bloodstained faeces, straining and scouring. Death can be quite rapid and is most common in young kids in the summer months. Your vet will advise a suitable dose of sulphamezathene as this drug is only obtainable under a veterinary prescription.

Infertility. This can be a problem in goats. One of the most obvious causes is lack of observation on the part of the goat keeper. The signs of heat in some goats are not obvious or they may be in season for only a very short period, e.g. a few hours. Learn to observe your goat very carefully. A good time to watch her is the half hour or so after milking when she has been fed and is not apparently doing much (see signs to look for on page 51). The animal which fails to come into season may be a so-called hermaphrodite in which parts of the genital tract have failed to develop. External signs need not be apparent in all those afflicted, and may only be diagnosed at post-mortem examination.

The lethargic, fat animal may fail to conceive. On the other hand one which is obviously undernourished may also fail to reproduce. In the latter, a good diet with access to an adequate mineral supplement may eventually solve the problem.

When buying a goat enquire into the breeding history of the family where known, and try to avoid the produce of a goat which kids infrequently.

Cloudburst. For no apparent reason, some goats produce all the symptoms of being in-kid whether they have been mated or not. Their weight increases, the udder develops and at about the time the kids are due they produce a quantity of pale fluid from the vulva. They should not be mated until the discharge has dried up. They may be milked normally.

Abortion. Goats can abort from rough handling, knocks, frights or infection. Violent exercise can also contribute to the premature loss of kids. Avoid feeding quantities of frozen greenstuff; this causes acute diarrhoea and sometimes results in abortion. If the goat *does* lose her kids prematurely, do watch out for signs of the afterbirth being retained: high temperature, loss of appetite and foul smelling vulval discharge. Call the veterinary surgeon immediately.

Milk fever. This is more likely to be found in the higher yielding older goat sometime before or after kidding. Symptoms may be glassy eyes with a dilated pupil and muscular weakness. If neglected the animal will collapse and die. Call the veterinary surgeon if the animal begins to stagger suspiciously. Bloat is frequently a symptom of milk fever and will resolve itself when the milk fever is treated, specific treatment for bloat under these circumstances is not to be undertaken. Do not let the animal lie on her side, but bolster her up with sacks of straw on either side so that she lies on her sternum. Avoid feeding concentrates for the next ten days to control the production of a high yield of milk too soon. It is essential that only about half the goat's production be withdrawn during the first fortnight. Check the newly kidded animal late at night for any signs as described above, and take immediate action if suspicions are aroused. By the morning it may be too late.

Mastitis. This is inflammation of the udder. It can be caused by injury and also rough handling by the kids or by the milker (gripping the actual udder as well as the teat when milking).

Learn to judge the normal warmth of the udder while milking. Any slight increase in temperature may indicate that mastitis is present. In severe cases, the milk may be clotted or stringy. The milk may not be pure white and can have a slightly 'off' or salty flavour. The udder may develop lumps and in severe cases will look shiny, red and be abnormally hard and painful.

Avoid feeding concentrates too soon after kidding, keeping to hay and a variety of tree leaves. The mastitis which develops after kidding may result in loss of appetite and the pupils of the eyes will appear as narrow slits.

Bathe the udder with hot fomentations to help relieve the congestion. Milk the udder frequently until the milk returns to normal, then gradually resume normal diet.

Varying degrees of mastitis exist from the mild to the severe and sometimes fatal.

The veterinary surgeon may advise administering an antibiotic by tube into the teat, after milking out the affected organ. The udder is then gently massaged to disperse the antibiotic. This dose is repeated once daily for three days, and the milk from the affected animal is best thrown away for a varying period depending on the preparation used. Milk from an animal so treated will not make yoghurt or cheese for about five days. Remember to wash hands carefully and thoroughly after handling a goat with mastitis. Should your goat develop mastitis, always handle it *after* you have finished handling the healthy ones.

Goat pox. This may occur in an animal that has been subjected to some stress such as a change of home or some kind of shock. It is catching, so milk the affected animal last. The udder develops pimples which gently weep and then form a crusty scab. It will clear up after a time. Dusting with boracic powder may help to dry the spots. The animal does not appear to be distressed at all by this complaint. The milk will be normal and safe to drink. It is possible for the milker to become infected by this virus and develop 'milker's nodules'.

There follows a list of some of the complaints to which animals of all ages may possibly succumb if their general management is not up to the mark in one or more ways.

Pneumonia. Poor housing, i.e. damp underfoot, draughty and cold, can predispose to pneumonia, especially where the goat is housed alone without the warmth of its fellows to rely on.

The symptoms to watch out for will initially be a faster rate of respiration than normal, which will immediately be apparent to the observant goat keeper. This may possibly be accompanied by groans. If she is standing, her back will be arched and her ears will hang unnaturally. She will look thoroughly miserable. Pneumonia strikes fast, and within a couple of hours she may be down on the ground unable to rise. By this time her breathing may appear bubbly. At the first signs consult the veterinary surgeon who may call to treat her. For complete recovery careful nursing is essential.

Give the animal a deep bed of straw, rug her warmly and prop her up on either side with straw bolsters. In really cold weather, adequately covered hot-water bottles will keep her warm. An infra-red lamp hung over her will help to maintain a comfortable temperature in the sick box.

Follow the vet's instructions carefully. Pneumonia calls for the best possible nursing during the long convalescence—possibly up to three months before complete recovery. She will need a laxative diet.

A milking goat will rarely recover her former milk production until the next lactation.

Entero-toxaemia. The first indication of this complaint will be a dead goat. It is most common amongst goatlings which have been fed a high proportion of concentrates. It is caused by an organism sometimes present in the gut which produces lethal toxins and thrives under these conditions. Goats can be vaccinated against it and should unexpected deaths occur, post-mortem examination and accurate diagnosis is essential in order to prevent further losses. It is the same disease as 'pulpy kidney' of sheep.

Over-eating. Gorging in the unlocked feed store, escape onto lush pastures, cabbages, kale, etc. and consequent overfilling with highly concentrated foodstuffs can lead to trouble. Excess fermentation in the large stomach will cause bloat, and symptoms of poisoning can result. Dullness, scouring, 'drunkenness' and, in extreme cases, prostration are symptoms. Immediate assistance is vital. Call the vet even *before the symptoms have developed.*

Bloat. This is another complaint caused by mismanagement. Any radical change in diet may cause bloat. The most likely cause is where animals are grazed on clovery grass in spring. It can be avoided by ensuring that all goats are fed an adequate ration of hay before being turned out to graze.

The animal which is blown will have a very tight flank on the left side caused by the excessive bacterial action in the rumen. This is often coupled with paralysis of the rumen. The distended organ may press on to the diaphragm inhibiting breathing, and in extreme cases will stop the heart from beating. If the animal is still on its feet when discovered, walk it around having administered a drench of half a pint of raw linseed oil or even cooking oil till it begins to 'burp' freely and eventually commences cudding. If the animal is down and still breathing, mark a point midway between the hip base and the last rib and four fingers below the back bone on the LEFT side. Pierce the animal with a sharp pointed knife. The foul smelling contents of the large stomach will spurt out. Keep the wound open by turning the knife until all the gas has escaped.

This is an easily avoidable condition. Herbal grazing and adequate browsing on a stomach lined with hay will help to prevent bloat. Adequate fencing to avoid the animal gaining access to clover leys, growing sugar beet or the cabbage patch will also help. Remember also always to secure food-store doors. Some animals do seem to be prone to bloat. In my experience it is the creature with the sulky nature, rather than the outward-going, friendly animal. A goat which does tend to bloat frequently is probably best not kept.

Diarrhoea. Most commonly this is a mechanical disorder caused through indiscreet feeding, heavy worm infestation or poisonous food. Too much lush grass or leguminous plants, access to frozen greenery and brassicas will cause diarrhoea to some extent. Allowing the goat access to oak leaves will help clear up diarrhoea caused by indiscreet or over-feeding. Treat the worm-infested animal with the appropriate worm remedy. The bottle-fed kid may develop diarrhoea through over-feeding, too-rapid feeding, or dirty utensils. Feed glucose in water for twenty-four hours (a tablespoon of glucose per pint of water) then gradually reintroduce milk in scrupulously clean bottles and avoid over-feeding. Kaolin powder offered to kids on a wet finger before the bottle feed will help to bind up an animal which may be a little loose.

Diarrhoea of the newly born kid is a separate problem. Frequently it arises in either the suckling kid or that fed on a bottle. Over-feeding is perhaps the commonest cause, although infection or such incidental things as change of the mother's diet or coming into season can contribute. The reserves of young kids can quickly become depleted and if the kid is obviously losing strength, professional help is indicated. First aid measures are diluting the mother's milk pint for pint with water to which is added one teaspoon of salt and one tablespoon of glucose. Feed this mixture four times daily, at blood heat, until improvement is evident. A dessertspoon of Kaolin powder mixed with water and given as a drink at least twice daily will help to bind the dung.

Tetanus. In counties where tetanus is known to exist it is worth having an anti-tetanus injection after a skin wound has been discovered. The symptoms of tetanus are an inability to swallow, a tense attitude with neck outstretched, muscular rigidity and bloat. Death will occur within nine days if the anti-serum has not been administered in time.

At this point, it is worth mentioning that combined

vaccines used for sheep against many of the commoner lethal diseases (tetanus is one) are also applicable in goats, and your vet will advise you as to their use.

This may appear to be a rather gloomy list of complaints, but most of them will be avoidable.

If a veterinary surgeon is indicated, do contact him early in the day if this is at all possible. Give him your name, address and telephone number and any symptoms you may have noticed. Clear instructions as to how your house may be reached will be greatly appreciated.

On arrival he will require a clean bucket of warm water, soap, a clean towel and a calm assistant to restrain the animal. Do follow his instructions as to the subsequent treatment the animal will need. Remember that a course of injections or medication in the feed or water may prove useless or even dangerous if not completed and administered at the correct interval of time and in the proper dosage. The vet will also want to wash his hands before leaving.

A well stocked medicine chest may save the life of a goat in an emergency—some suggestions follow:

Savlon, Acriflex,
Dettol or other disinfectant
Boracic powder
Golden Eye ointment
Kaolin powder
Bicarbonate of soda
Linseed oil
Epsom salts
Worm powders—Thibenzole
Cotton wool
Scent-free soap
Teaspoon

Measuring jug
Bowl and clean towel
Clinical thermometer
Scissors
Drenching bottle
Rug

This box needs to be stored in a dry, goat-proof shed, possibly on a high shelf in the food store where it is easily accessible when needed.

Diseases due to mineral deficiencies are remotely possible and may be suspected when the goat's behaviour varies from the norm. They really require a skilled diagnosis coupled with a knowledge of the locality. Your local vet is in the best position to offer advice in this respect.

8 General Management

If the goat is deprived of the company of her own kind, don't be surprised if she adopts you and the family as her herd. This may sound charming, if not coy, but in reality it can become rather tedious. She will bleat when you are not with her in an attempt to maintain communication. Tea on the lawn can be riotous, especially if her kids are about as well. We have on more than one occasion had to fence ourselves in on the lawn with a circle of pig netting to preserve some semblance of polite society at tea-time.

If you are lucky enough to breed a nanny kid you may well be tempted to keep her as a possible eventual replacement for your goat. When old enough she will be able to supply milk while her dam is dry. This situation will, to a certain extent, relieve you of the responsibility of being companion *and* herd leader *all* the time. If stall fed, mother and daughter can be housed where they can see and hear each other if not actually stalled together. When the weather is suitable they can be tethered outside within sight of each other. Animals kept loose in the garden or a paddock will graze companionably while the solitary goat will tend to wait at the gate all day for attention.

If it is convenient to keep only one goat and you like to have her loose sometimes, a half door fitted at the external kitchen entrance will mean that she can still run to you for comfort as she will be able to see and hear you. You will be relieved of the responsibility of having an assistant in the kitchen, as she will invariably overstay her welcome and may do rather more than consume the contents of the compost bucket.

Our first responsibility to the goat and the family is to establish the human position as herd leader. She will soon learn who feeds her, milks her, loves her; she will obey this person regardless of others. She will leave the tastiest morsel on hearing her milker call, even when it is *not* milking time. Plenty of handling, reassurance and firmness will establish this. No one wants a goat which rules the household with her capricious wiles, and firm handling from the start will quickly establish the correct relationship. Diffident, indeterminate and unsure handling will quickly put the goat in the ascendancy.

We have all heard of the goat who won't be milked. Approach the goat with firmness. Have someone ready to hold her if she *does* move. Better still, try milking her in a yoke with food for her to eat to keep her still even if only for a few minutes. When she realises that her milker is stronger, both mentally and physically, she will soon acquiesce and stand still. But

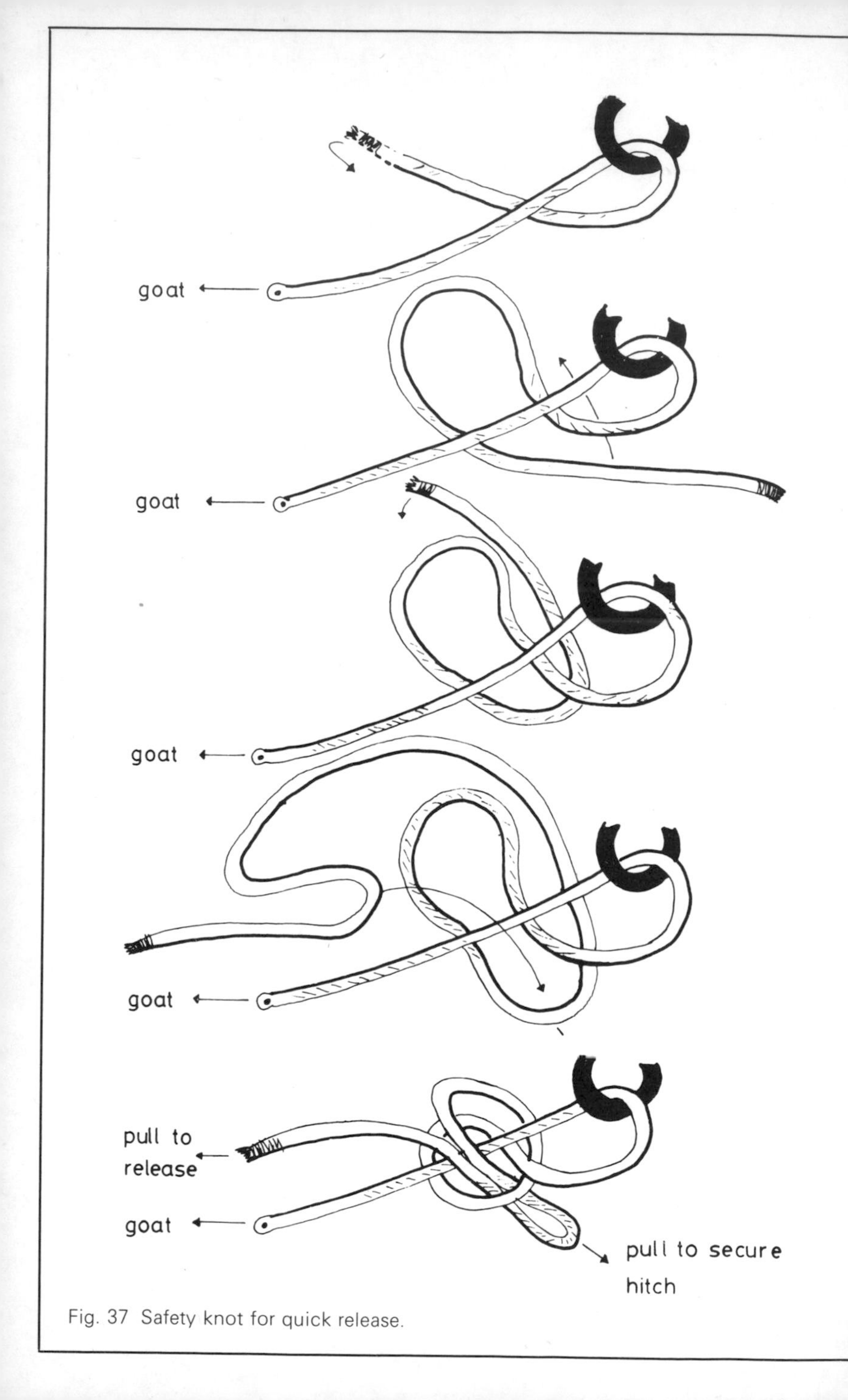

Fig. 37 Safety knot for quick release.

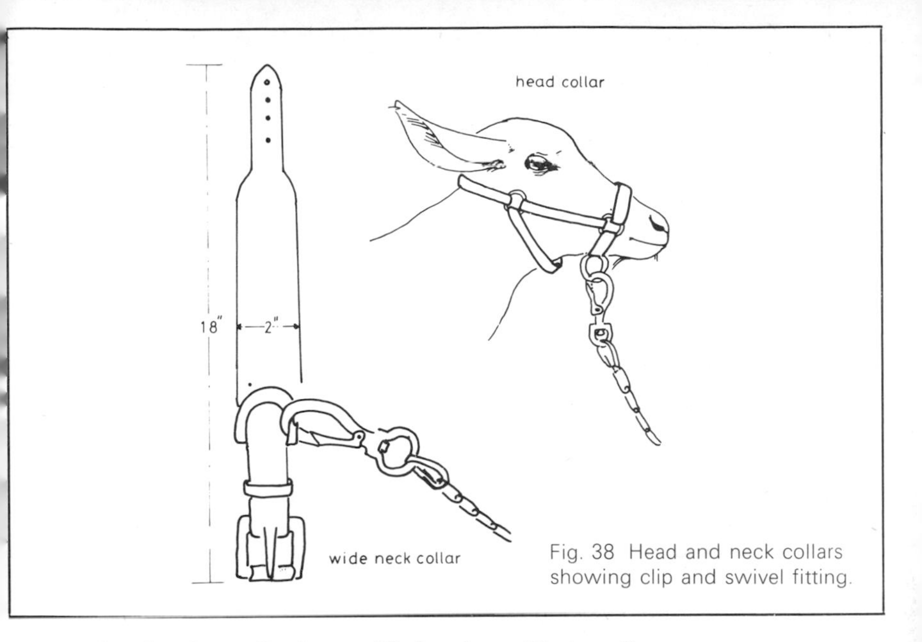

Fig. 38 Head and neck collars showing clip and swivel fitting.

do check to find out if she is suffering from some genuine discomfort which may make her fractious. She will be relieved to be rid of the milk in her udder and will soon learn to stand quite still throughout milking. Unsure handling will worry her.

Whenever an animal has to be tied it is a good idea to use the safety knot shown in Fig. 37. The particular advantage of this knot is that it can be quickly unfastened by simply pulling one of the strings.

The month-old kid is not too young to have a collar on once a day, and during this time she can have her first lessons in leading. A wide collar which does not cut into her neck, and a short, stout lead will be suitable (see Fig. 38). A head collar could also be used, but these are not so readily available. A small goat or sheep-bell can be attached to her collar and this will enable her to be heard if she's loose in the garden. On the other hand, if several goats wear bells one soon learns to judge what they are doing by the 'music'; trouble can be quickly dealt with if the bells start to ring furiously, for example if the goats are being involved in a chase. Dead silence might also be ominous as it may mean that the goats have raided the chicken food.

A goat can be taught to be led, and she will also follow her handler if loose. She does not, however, appreciate being driven. If chased the goat takes avoiding action by jumping to one side then facing the pursuer. This is unlike the cow, horse and sheep which tend to flee from danger, relying on speed for escape.

When teaching the kid to lead, stand at her near shoulder with the lead in the left hand, as shown in Fig. 39. Hold her round the rump with the right arm pushing her forward saying, 'Walk on' firmly. Just one step forward is sufficient at first. Make a fuss of her and turn her loose once more. The next day repeat the process and she will soon get the idea and need less and less pressure from behind. Don't, under any circumstances, pull her from the front; her reaction will be to brace all four legs and stand stock still. She may even collapse. I've seen goats pulled along (in the

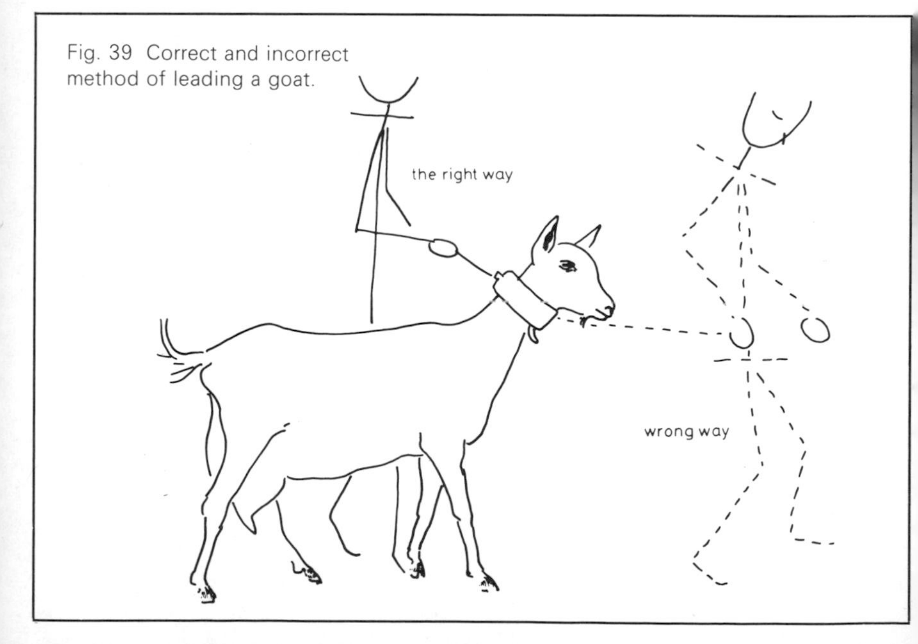

Fig. 39 Correct and incorrect method of leading a goat.

show ring even) and collapse as the collar was pulled too tight in the wrong place. The kid will want to play but she must learn that the collar means *work*. She *must* be obedient while she's in harness.

The kid will not be tethered till she is at least nine months old, and this will best be done for short periods at first, especially if she's on her own. Left alone she will bleat piteously. Stay with her at first and then creep away as her attention is diverted by a succulent morsel. She will soon learn to accept this mode of life, especially if she is tethered near to another goat. She will soon indicate if the place is not to her liking. A poor tethering spot will be indicated by the goat pulling on her tether and bleating but when moved to a better patch she will soon settle to browse and graze. Always remember that all goats need to be taken under cover as soon as it rains, and, if tethered they also need to be moved several times a day.

Where a goat is kept on soft litter there will be little chance of exercise to wear away the under surface of the foot. The hoof will grow too long, turning under to form a crevice between the dry horn and the foot sole; it may even develop a clog-like appearance as shown in Fig. 40.

The resulting long hoof can become deformed and lameness will result. Regular trimming, at least once a month, will prevent this. The best tool for the job is a sharp hoof-paring knife obtainable from a dairy stockist or saddler. A kitchen paring knife will do if it is really sharp but has a blunt blade end (see Fig. 40). The animal must be securely held by an assistant or tied against a wall. The operator stands at the goat's shoulder facing her tail, and holds the foot firmly in one hand progressively cutting back the hoof working from the heel to the toe till it is pared level with the sole. The heel has a certain amount of dry material

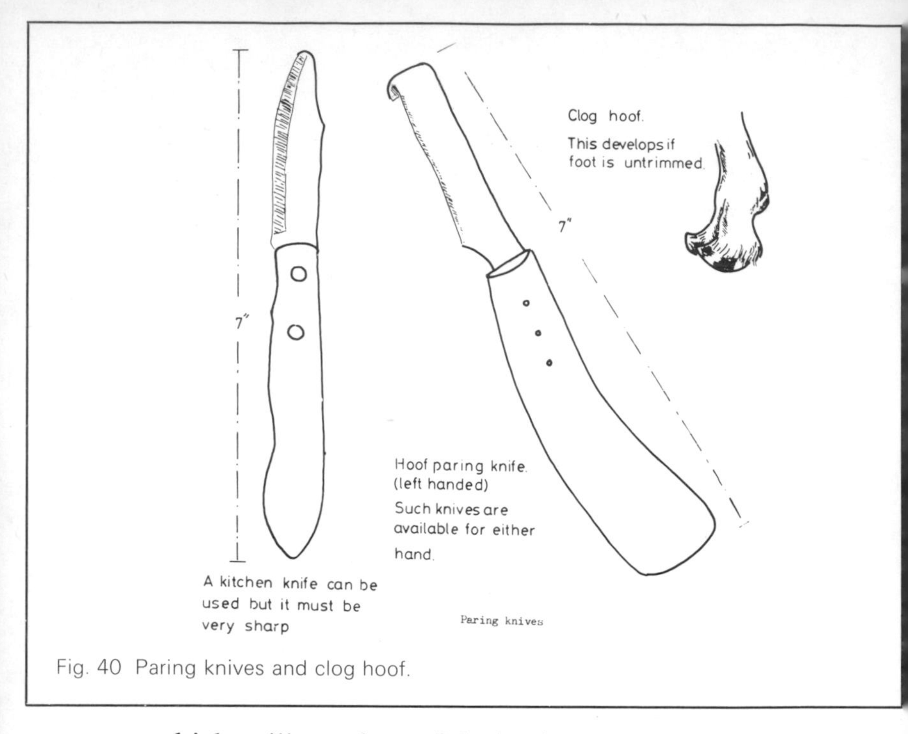

Fig. 40 Paring knives and clog hoof.

which will need careful shaping to avoid drawing blood. The goat with neglected hooves cannot be completely trimmed in one operation. Pare small quantities of the horny growth weekly allowing her plenty of exercise on firm ground to harden the feet. The newcomer to goat keeping is advised to watch an expert first.

When the kid is about six weeks old she may well need her hooves trimmed. From birth it is wise to pick her hooves up one by one getting a member of the family to help to hold her still and reassure her. Obviously the kid will struggle at first, but the animal which has been handled freely and firmly from birth will learn to accept the necessary indignity of hoof trimming. A rock or block of concrete, roughly 18 inches square and a foot high, kept in her stall will help to maintain hard, healthy hooves.

It is useful to have a dandy-brush available. The

animal which spends a lot of time out of doors is best not brushed too vigorously especially in the winter when too much protective wool and valuable grease may be removed. In the summer, however, and when she is moulting, a few brushes followed by a chamois leather will greatly improve her appearance. The moulting goat may go off her food slightly for a few days till the new coat comes through.

When considering straw for litter, there are several types available. Wheat straw is ideal as it does not pap down into a damp mass, being of a fairly rigid nature. Barley straw is suitable, but generally more expensive especially if it is spring sown as it can also be used for animal feed. Oat straw is also used for feed and makes a reasonable litter. Wheat straw is, then, generally cheaper but is more often only available in the heavier, more fertile, areas of the country.

Wood shavings are sometimes free for the collection and sugar beet pulp sacks are ideal for holding this. Shavings and coarser sawdusts from soft-wood timber are generally more suitable than those from the hardwoods.

Bracken can be gathered after it has died down if you live near common land, but it is extremely sharp to handle so wear leather gardening gloves to prevent the splinters from shredding your fingers and use a scythe or hook to cut it down. Peat moss is very dusty even if the coarsest grade is available, but it is very economical as it goes a long way and is very absorbent of both urine and smells.

Goats can either be cleaned out daily or a deep litter built up. In the latter case cleaning out will only be needed at six monthly intervals. Additional litter can be added from time to time depending on the weather conditions. You will have to dispose of the litter whichever method is favoured. It is an ideal manure for the garden after being composted for a time. To make a manure heap, mark out a plot no smaller than 6 feet × 4 feet at a sufficient distance from the milking area to keep the fly problem down. Build the heap carefully, paying particular attention to the sides which should be nearly vertical. If the material seems at all dry, a bucket of water sprinkled on will help the bacteria to start work on the litter. Properly made, the resulting compost will be friable and rich in plant food. I like to cover the heap with plastic fertilizer bags to prevent the rain from leeching these nutrients away. A well-built manure heap will gladden the eye of any gardener and may be ready for use in about six months or less depending on the time of year; the heap made in summer being ready in a shorter time. I usually have two or three heaps at the same time. One heap will consist of fresh manure, the second will have been watered and turned top-to-bottom and sides-to-middle to assist the composting process, while the third will possibly be rotted and ready for use. The litter from the deep litter house will need to be shaken out well, and possibly even watered, as it is put on to the heap.

Specialised tools are not really needed for the one-goat family as existing gardening gear can be used quite easily. A garden fork will be adequate for littering and cleaning out the goat house, on the other hand, the fork with round, rather than square tines will make the job easier. Remember that excreta are corrosive and well-greased tines will speed the job up enormously as the litter slips on and off more easily. A two-tincd hay fork is ideal for shaking up litter, hay making and gathering up hedge trimmings. These can usually be acquired for a song at farm sales.

Secateurs and garden shears (particularly those with a notched blade near the pivot) are ideal for branch clipping and will speed up the gathering of browsings. As these browsings are removed from the goat shed, they can be tied in bundles and used for pea and bean sticks. Elm, besides being very palatable to goats, forms an ideal pea stick.

A scythe is most useful for cutting grass and weeds for hay but some practice may be necessary for satisfactory operation. Remember that it must be really sharp and a good 'stone' will be needed to maintain an effective edge during use.

A plastic-bristled yard broom will last far longer than the usual variety, even on concrete. Kleeneze make a really good one which will last for years if it is not left out in the rain. A couple of plastic-bristled scrubbing brushes are essential.

Some form of wheel barrow will lighten the burden of carting litter, manure, food, etc. Double-axled carts will carry more but are obviously more expensive.

Finally, last but not by any means least, some form of records are needed, however simple.

A simple file will hold the registered papers of pedigree animals, records of matings, etc. This will keep all these particulars in one place.

A postcard on the meal store wall, stating the date a quantity of food was bought and the date it was finished, will give an idea of how much concentrates really *are* being fed. It is easy to say, 'Oh, she only has a handful twice a day', and then to find that more is being bought than was expected. Are the children perhaps depleting the goat's ration by feeding it unofficially to the guinea pigs?

Information recording every-day occurrences —the dates the goat's seasons are noticed, when she is due to go to the billy, the date the kids were first noticed to be moving, the appropriate time she is due to be dried off, changes in rations, any slight digestive upsets, if she is off her food, etc.—can go on a card on the wall or in a Boot's Scribbling Diary. It is

Feed record

Date purchased	Feed type	Quantity	Cost/cwt	Date finished

Breeding record

Name of nanny	Date served	Name of sire	Date due	Date kidded	Number & sex of kids

Milk record

Name of goat:		
Date a.m. ¦ p.m.	Quantity	Comments

Sales record

Name of goat	Date sold	Purchaser	Price

Fig. 41 Record cards.

very easy to think that all these things can be remembered, but we are all fallible. It only needs someone in the family to be ill and a possible visit to the billy is forgotten, resulting in a later kidding than one had hoped. Had the date been recorded in a prominent place, this vital appointment/assignation need not have been missed. I like to keep a particular note of each animal's pre-kidding behaviour and then I have more idea of what to expect in subsequent kiddings. These can then be filed indoors with each animal's record. A list of useful telephone numbers can also be kept indoors, such as those of the vet, miller, local goat-club secretary, etc. Suitable layouts for record cards are shown in Figs. 41 and 42.

Telephone numbers

Service	Address	Exchange / std no	Number
Billy owner			
Dairy stockist			
Feed merchant			
Goat Club sec.			
Vet			

Fig. 42 Record card for useful telephone numbers.

A 25 pound spring balance hung in the meal store will be useful for recording daily milk production. It will also be useful for weighing the ration of concentrates and the hay nets.

These records may well prove invaluable and can even help a prospective buyer of any stock you may have for sale.

In most parts of the country there are local goat clubs whose members will welcome and give advice to a newcomer. In fact my experience is that they are only too willing to help in any way at all. These clubs very often run local goat shows where members meet for a talk and compare their stock. These shows are run on a very friendly basis. Classes for most breeds of goats and of all ages, and also for the non-pedigree animals, are held. Milking competitions are also held for animals of unknown as well as known ancestry.

The British Goat Society has many booklets and leaflets available and is the headquarters to which any registration papers should be sent. A registration card does *not* come with a goat when it changes hands. The seller is responsible for sending the animal's card and a completed transfer form to the British Goat Society with the relevant fee. The society will then send the card to the new owner, having registered the change. Membership of the society amongst other benefits entitles one to receive the monthly journal and year book. The address is:

British Goat Society
Rougham
Bury St Edmunds
Suffolk IP30 9LJ

Goats and the Law

Nuisance. The owner of a goat must ensure that his animal does not cause a nuisance to neighbours in any way. 'Nuisance' can cover anything from noisiness to unpleasant smells!

Movements of Animals (Records) Order 1960. Goat keepers are obliged to keep a record of all movements of their goats, including sales and purchases, mating and slaughter journeys. Your local Ministry of Agriculture office can provide further details.

Finally, good luck with your goat. Remember she is an individual; observantly, firmly and lovingly handled she will give much pleasure and profit, presenting you with household quantities of dairy produce and meat for the minimum of outlay.

Appendix—Useful Information

Further Reading

Breeds of Goats
Goat Keeping
Goat Feeding
Dairy Work for Goat Keepers
— all obtainable from the British Goat Society

Goat Husbandry, David Mackenzie, Faber & Faber.
Exhibition and Practical Goat Keeping, Joan Shields.
All About Goats, Lois Hetherington, Farming Press.
The Goat Keeper's Guide, Jill Salmon, David & Charles.
The Backyard Dairy Book, Len Street and Andrew Singer, Prism Press.
Herbal Handbook for Farm and Stable, Juliette Bairacli Levi, Faber & Faber.
British Poisonous Plants, HMSO Bulletin No. 161.
Electric Fencing, HMSO Bulletin No. 147.

Farming Organically
Self-Sufficient Smallholdings
Smallholder's Harvest
Looking at Livestock
— all obtainable from The Soil Association, Walnut Tree Manor, Haughley, Stowmarket, Suffolk IP14 3RS.

Suppliers of Equipment

Self-Sufficiency Supplies Ltd, Priory Road, Wells, Somerset.
Small Scale Supplies, Widdington, Saffron Walden, Essex CB11 3SP.

Index